SH 328 .O24 2008
Oceans past

MHCC WITHDRAWN

D1774362

Oceans Past

Oceans Past

Management Insights from the History of Marine Animal Populations

*Edited by David J. Starkey,
Poul Holm
and
Michaela Barnard*

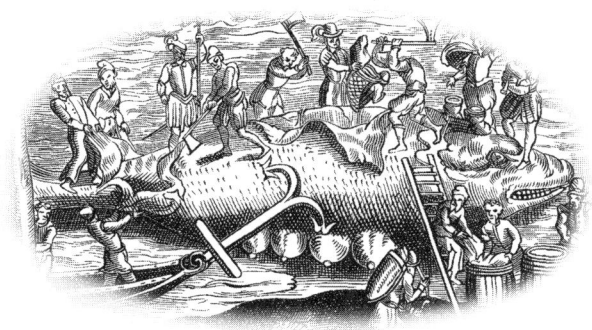

London • Sterling, VA

First published by Earthscan in the UK and USA in 2008

Copyright © David J. Starkey, 2008

All rights reserved

ISBN: 978-1-84407-527-0

Typeset by JS Typesetting Ltd, Porthcawl, Mid Glamorgan
Printed and bound in the UK by TJ International, Padstow
Cover design by Susanne Harris

For a full list of publications please contact:

Earthscan
8–12 Camden High Street
London, NW1 0JH, UK
Tel: +44 (0)20 7387 8558
Fax: +44 (0)20 7387 8998
Email: earthinfo@earthscan.co.uk
Web: **www.earthscan.co.uk**

22883 Quicksilver Drive, Sterling, VA 20166-2012, USA

Earthscan publishes in association with the International Institute for Environment and Development

A catalogue record for this book is available from the British Library

Library of Congress Cataloging-in-Publication Data

Oceans past : management insights from the history of marine animal populations / edited by David J Starkey, Poul Holm, and Michaela Barnard.
    p. cm.
  ISBN-13: 978-1-84407-527-0 (hardback)
  1. Fishery management. 2. Whaling--Management. 3. Marine resources--Management. I. Starkey, David J. (David John), 1954- II. Holm, Poul. III. Barnard, Michaela.
  SH328.O24 2008
  333.95'6--dc22
                                      2007034788

The paper used for this book is FSC-certified and totally chlorine-free. FSC (the Forest Stewardship Council) is an international network to promote responsible management of the world's forests.

Contents

List of Figures and Tables	*vii*
Acknowledgements	*xi*
List of Contributors	*xiii*
List of Acronyms and Abbreviations	*xvii*

Foreword: Future Knowledge of Life in Oceans Past — xix
Jesse H. Ausubel

1 Oceans Past: History Meets Marine Science — 1
   David J. Starkey, Poul Holm and Michaela Barnard

2 Invasive or Native? The Case History of the Common Periwinkle Snail (*Littorina littorea*) in Northeast North America — 7
   April M. H. Blakeslee

3 Social Conflict, Over-Fishing and Disease in the Florida Sponge Fishery, 1849–1939 — 25
   Loren McClenachan

4 'Quite the Choicest Protein Dish': The Costs of Consuming Seafood in American Restaurants, 1850–2006 — 47
   Glenn A. Jones

5 Baiting Our Memories: The Impact of Offshore Technology Change on Inshore Species Around Cape Cod, 1860–1895 — 77
   Matthew G. McKenzie

6 Mapping Historic Fishing Grounds in the Gulf of Maine and Northwest Atlantic Ocean — 91
   Stefan Claesson

7 There She Blew! Yankee Sperm Whaling Grounds, 1760–1920   109
   *John L. Bannister, Elizabeth A. Josephson, Randall R. Reeves and
   Tim D. Smith*

8 Depletion within a Decade: The American 19th-Century
   North Pacific Right Whale Fishery   133
   Elizabeth A. Josephson, Tim D. Smith and Randall R. Reeves

9 Sperm Whale Catches and Encounter Rates during the 19th
   and 20th Centuries: An Apparent Paradox   149
   *Tim D. Smith, Randall R. Reeves, Elizabeth A. Josephson,
   Judith N. Lund and Hal Whitehead*

10 Understanding the Dynamics of Fisheries and Fish Populations:
   Historical Approaches from the 19th Century   175
   Julia Lajus

11 A Political History of Maximum Sustained Yield, 1945–1955   189
   Carmel Finley

Afterword: Lost and Found in the Past   207
Paul E. Waggoner

Index   215

List of Figures and Tables

FIGURES

2.1	Rarefaction curves for haplotype diversity and sampling effort	16
2.2	Expected number of haplotypes in Europe versus North America per sampling effort from the Wares et al (2002) haplotype dataset	17
3.1	Schooners at Key West Sponge Docks, 1902	29
3.2	Sponge workers trimming sponges, Nassau, the Bahamas, circa 1932	30
3.3	Sponges drying on a platform at the rear of sponge house, Key West, December 1921	31
3.4	Florida sponging grounds	34
3.5	Sponge catches, Key West and Tarpon Springs, 1888–1918	35
3.6	Greek sponge divers, Tarpon Springs, circa 1931	37
3.7	Florida sponge catches, 1839–1940	42
4.1	Bill of fare from Jones' Hotel (June 1845)	49
4.2	Bill of fare from a banquet celebrating the opening of the New York and Canada railroad (November 1875)	50
4.3	Bill of fare from Keeler's European Hotel (March 1897)	52
4.4	Bill of fare from Union Oyster House (November 1920)	53
4.5	The Fishermen's Grotto menu (July 1957)	54
4.6	The Consumer Price Index and its pre-1913 equivalent	57
4.7	Hypothetical temporal price behaviour for item following: (a) CPI; (b) CPI changing to CPI + 2 per cent after 1950; (c) CPI changing to CPI − 2 per cent after 1950	58
4.8	Comparison of the CPI, current dollar and constant dollar wholesale price of live lobster derived from the NOAA fisheries database	60
4.9	Inflation-adjusted price of canvas-back duck, 1860–1910, and abalone, 1920–2004	65
4.10	Postcard of abalone drying on the shore	68
4.11	Postcard of abalone shells lying outside a processing plant in Santa Barbara, California	69
4.12	Price series, American lobster, 1885–2004	72

6.1	The study area of the Gulf of Maine Cod Project	92
6.2	Chart No 8: Fishing grounds from Penobscot, Maine, to Cape Ann, Massachusetts	95
6.3	Coastal banks and inshore grounds of the Gulf of Maine, Monhegan to Cape Cod	97
6.4	A page from the logbook of the Beverly, Massachusetts, fishing schooner *Dove* (1854)	98
6.5	Fishing charts of known and exploited fishing grounds and banks in the Gulf of Maine during the late 19th and early 20th centuries	101
6.6	Goode Fishing Chart No 5: Fishing Grounds of the Gulf of St Lawrence (1887)	102
6.7	Sailing courses of 27 US schooners from Beverly and Salem, Massachusetts, that fished on the Scotian Shelf during the 1850s	103
6.8	Seasonal fishing by US vessels on the Scotian Shelf showing a concentration of effort on Banquereau and Sable Island Banks	104
7.1	Townsend's sperm whale chart, Pacific section	112
7.2	Consolidated sperm whaling locations in (a) the Atlantic and Indian Ocean and (b) the Pacific Ocean	118
7.3	Seasonal distribution of sperm whaling grounds in the Pacific Ocean	119
7.4	An example of Townsend's sperm whale plots (Pacific Ocean)	130
8.1	Right whale encounters in the *Maury Abstracts* (dots) and regions selected for analysis	135
8.2	(a–o) Daily positions of whaling vessels in the *Maury Abstracts*, pre-1839 to 1852	137
8.3	Numbers of US vessels whaling in the North Pacific, 1839–1853	142
8.4	Average catch reported by the *Whalemen's Shipping List and Merchants' Transcript* for the North Pacific fishery	143
8.5	Right whale encounter rates, 1839–1851, with 95 per cent confidence intervals: (a) Gulf of Alaska region; (b) Kamchatka region; (c) Kurile Island region; (d) Sea of Japan region; (e) offshore region	144
9.1	Estimated annual global removals of sperm whales, circa 1750–2000, and encounter rates in the mid-19th century	150
9.2	Estimated annual removals of sperm whales, circa 1750–2000, for North Pacific and South Pacific	160
9.3	The locations of 20th-century catches shown by shaded 5° squares of latitude and longitude	161
9.4	The annual proportion of whaling days with encounters of sperm whales on the Japan Ground, 1822–1852	162
9.5	The proportion of whaling days with encounters of sperm whales in areas adjacent to the Japan Ground, 1822–1852	164
11.1	US High Seas Fisheries Policy	194

TABLES

2.1	Evidence relating to the cryptogenic status of *Littorina littorea* in northeast North America	11
3.1	Species targeted in the Florida sponge fishery	28
4.1	Prices charged for selected seafood items by Bristol's Oyster and Chop House, New York City, January 1900	61
7.1	Numbers of sperm whaling locations identified by ocean basin and author	113
7.2	Numbers of Yankee sperm whaling grounds and locations identified by ocean basin	115
7.3	Consolidated list of Yankee sperm whaling locations	116
7.4	Seasonality of visitation, by location and source	120
7.5	Main periods of occupation of grounds by ocean basin	124
9.1	Four assumptions underlying the apparent paradox revealed by three analyses of 19th- and 20th-century sperm whaling	151
9.2	Information regarding 18th- and 19th-century American open-boat whaling voyages derived from published sources	155
9.3	Aggregate sperm oil imports to US ports and voyage-based oil production	155
9.4	Average oil yield of sperm whales reported in 19th-century logbooks for voyages by vessels of different types	157
9.5	Estimated number of sperm whales removed	159
9.6	Sperm oil landed and freighted for voyages ending in 1859	168
9.7	Sperm oil freighted and the number of shipments arriving in New England in 1859	169
9.8	Encounter rates of different clans of sperm whales for whaling vessels moving at different speeds	170

Acknowledgements

The editors are grateful to many individuals and institutions for their contributions to the production of this volume. *Oceans Past* had its genesis in an international conference held under the auspices of the History of Marine Animal Populations (HMAP) project in Kolding, Denmark, in October 2005. Our thanks are due to the committee that organized the academic programme of this meeting (Poul Holm, chair, Andrew A. Rosenberg, Tim D. Smith and David J. Starkey) and to Anne Husum Marboe for her sterling work in solving the countless administrative, practical and logistical problems that invariably arise in the mounting of such events. We are pleased to acknowledge the generous sponsorship provided in support of the conference by SAS (Scandinavian Airlines), MarBEF (Marine Biodiversity and Ecosystem Functioning – EU-funded Network of Excellence), NORMA (Nordic Research Network in Maritime Studies) and the Danish Research Council for the Humanities. We would also like to record our appreciation of the long-standing support afforded by the Alfred P. Sloan Foundation – and the inspiration of its programme director, Jesse H. Ausubel – to the Census of Marine Life and HMAP projects, in general, and to the publication of this volume in particular. Debts of gratitude are owed to the 110 researchers who participated in the Oceans Past conference, especially to those who subsequently revised their papers for presentation in this volume, and to the referees who provided highly valuable advice to these authors in the initial stages of the editorial process. Lastly, we offer thanks to Hamish Ironside, Michael Fell, Alison Kuznets and their colleagues at Earthscan for their efficiency and expertise in bringing this book to fruition.

David J. Starkey, Poul Holm and Michaela Barnard
July 2007

List of Contributors

Jesse H. Ausubel helped establish the Census of Marine Life research programme. He is a programme director for the Alfred P. Sloan Foundation and a faculty member at The Rockefeller University in New York City, where his work examines long-term interactions between technological and environmental change.

John L. Bannister is a cetacean biologist and fellow of the Zoological Society of London, the Linnean Society and the Western Australian Museum. He has served as junior whale fishery inspector in South Georgia, research scientist at the UK National Institute of Oceanography and at Australia's Commonwealth Scientific and Industrial Research Organization (CSIRO), director of the Western Australian Museum and chair of the Scientific Committee of the International Whaling Commission.

Michaela Barnard is research fellow at the Maritime Historical Studies Centre, University of Hull, UK. As well as managing the HMAP Data Facility, she teaches maritime and marine environmental history and undertakes research into the business dimensions of fisheries and shipping history.

April M. H. Blakeslee is a marine ecologist at the Smithsonian Environmental Research Center. She was awarded a PhD in May 2007 by the University of New Hampshire for her research into marine invasions, focusing on population genetics and parasites.

Stefan Claesson is a research associate in historical ecology and maritime archaeology at the Ocean Process Analysis Laboratory, University of New Hampshire. His research focuses on the sustainable development of cultural heritage, with particular reference to maritime archaeological resources in the Gulf of Maine.

Carmel Finley is a historian of science. She graduated from the University of California, San Diego, and is revising her dissertation, 'Fish, fisheries science and foreign policy, 1920–1960', for publication.

Poul Holm is rector, and professor of environmental history, at Roskilde University, Denmark. His research focuses on the role of humans in the long-term development of marine environments. He is chair of the HMAP Executive Committee, and has been a member of the Census of Marine Life Scientific Steering Committee since 2000.

Glenn A. Jones earned a PhD in marine geology from Columbia University. Having served as research scientist at the Woods Hole Oceanographic Institution, he is professor of oceanography at Texas A&M University at Galveston, where he teaches courses on marine geology, oceanography, and ocean and coastal resources.

Elizabeth A. Josephson is a researcher at the National Oceanic and Atmospheric Administration in Woods Hole, MA, where she has been integral to the History of Marine Animal Populations project. She specializes in interpreting logbooks from American open-boat whaling voyages, and oversees the extraction and curation of logbook data.

Julia Lajus is director of the Center for Environmental and Technological History, European University at St Petersburg, Russia, and senior researcher at St Petersburg Branch of the Institute for the History of Science and Technology. She has co-ordinated the HMAP White and Barents Sea project since 2001 and is a member of HMAP Steering Group.

Judith N. Lund is a historian who focuses on the American whaling industry and local history of the New Bedford area. A former curator of the New Bedford Whaling Museum, she now works from her home in Dartmouth, Massachusetts.

Loren McClenachan is a PhD candidate at the Center for Marine Biodiversity and Conservation at the Scripps Institution of Oceanography. Her research addresses long-term ecological change and the effects of historic fishing on Caribbean coral reef ecosystems, with a focus on the Florida Keys, the Bahamas and Cuba.

Matthew G. McKenzie was awarded a PhD by the University of New Hampshire in 2003. Having taught at sea with the Sea Education Association, he is currently an assistant professor of history at the University of Connecticut.

Randall R. Reeves is chairman of the World Conservation Union (IUCN) Cetacean Specialist Group and is involved in a variety of whale and dolphin conservation projects around the world. His participation in the HMAP World Whaling Project stems from a long-standing interest in whaling history and its implications for conservation.

Tim D. Smith is a fisheries biologist who specializes in international fisheries research and management, especially relating to marine mammals and the history of fisheries science. He recently retired from the National Oceanographic and Atmospheric Administration and continues to direct the HMAP World Whaling Project. He is adjunct professor at the University of New Hampshire.

David J. Starkey is reader in maritime history, and director of the Maritime Historical Studies Centre, at the University of Hull, UK. His research interests embrace all aspects of past interactions of humans and the marine environment, especially fisheries and shipping history. He has been a member of the HMAP Steering Committee since 2000.

Paul E. Waggoner was director of the Connecticut Agricultural Experiment Station in New Haven, and continues there as a distinguished scientist. Recently he has studied land use and the environmental impact of consumption. He participated in the preparation of the Census of Marine Life Baseline Report and Annual Highlights.

Hal Whitehead is a university research professor in the department of biology at Dalhousie University. His research focuses on social organization and cultural transmission in the deep-water whales, especially sperm whales, but he also works on their ecology, population biology and conservation.

List of Acronyms and Abbreviations

AFLP	amplified fragment length polymorphism
ATA	American Tuna Association
bbl	barrel
CBD	Centre for Biological Diversity
CCCC	Cape Cod Community College
CFR	*Commercial Fisheries Review*
COADS	Comprehensive Ocean–Atmosphere Dataset
CoML	Census of Marine Life
CPI	Consumer Price Index
CPUE	catch per unit effort
CSDGM	Content Standards for Digital Geospatial Metadata
DCR	Daily Catch Record
DNA	deoxyribonucleic acid
dpi	dots per inch
FAO	Food and Agriculture Organization of the United Nations
FGDC	Federal Geographic Data Committee
FMAP	Future of Marine Animal Populations programme
FOR	*The Fish and Oyster Reporter*
ft	international foot (1ft = 12 inches [in] = 30.48 centimetres [cm])
FTR/HF	Falmouth Town Records, *Herring Fishery* (Falmouth, Massachusetts)
FTR/m	Falmouth Town Records, microfilms (Falmouth, Massachusetts)
FTU	*Florida Times Union*
GIS	geographic information system
GMCP	Gulf of Maine Cod Project
HMAP	History of Marine Animal Populations
HTR	Harwich Town Records (Harwich, Massachusetts)
IATTC	Inter-American Tropical Tuna Commission
ICNAF	International Commission for North Atlantic Fisheries
IM	isolation with migration analysis
in	international inch (1in = 1/12 foot [ft] = 2.54 centimetres [cm])
ITS	internal transcribed region
IWC	International Whaling Commission
kg	kilogram
KWC	*Key West Citizen*
KWCHR	Key West Custom House Records

kya	thousand years ago
lb	avoirdupois pound (1lb = 16 ounces [oz] = 453.59 grams [g])
MarBEF	Marine Biodiversity and Ecosystem Functioning
MCIF	Massachusetts Commissioners of Inland Fisheries
m	metre
mm	millimetre
MSC	Massachusetts, Secretary of the Commonwealth
MSY	maximum sustained yield
mt	metric tonnes
n	total population sample size
NARA	National Archives Records Administration
NARA/RR/W	National Archives Records Administration, Regional Repository, Waltham (Massachusetts)
nm	nautical miles
NMFS	National Marine Fisheries Service (NOAA)
NRC	National Research Council (Washington, DC)
NOAA	National Oceanic and Atmospheric Administration
NORMA	Nordic Research Network in Maritime Studies
OBIS	Ocean Biogeographic Information System
OPA	Office of Price Administration
oz	avoirdupois ounce (1oz = 1/16 pound [lb] = 28.35 grams [g])
PC	paradox coefficient
PEM	Peabody Essex Museum (Salem, Massachusetts)
PFA RAN	St Petersburg Branch of the Archives of the Russian Academy of Science
PMM	Pilgrim Monument Museum (Provincetown, Massachusetts)
RTR	Rochester Town Records (Rochester, Massachusetts)
SAS	Scandinavian Airlines
SCUBA	self-contained underwater breathing apparatus
SIOA	Scripps Institution of Oceanography Archives (University of California, San Diego)
STR	Sandwich Town Records (Sandwich, Massachusetts)
TOPP	Tagging of Pacific Predators
UK	United Kingdom
UN	United Nations
UNH	University of New Hampshire
US	United States
USDCL/BF	United States Department of Commerce and Labour, Bureau of Fisheries
USFC	United States Fish Commission
USFWS	United States Fish and Wildlife Service
USSR	Union of Soviet Socialist Republics
WGS84	1984 World Geodetic System
WSL	*Whalemen's Shipping List and Merchants' Transcript*
VMAP	vector map level 0
ypb	years before present

Foreword

Future Knowledge of Life in Oceans Past

Jesse H. Ausubel

An English prelude

I am always comfortable and happy among historians because my father, whom I liked very much, was a historian. He specialized in 19th-century Britain. Every few years, my father would bring the family by transatlantic passenger liner from New York City to Southampton for a long stay to conduct his research. I crossed the North Atlantic by ship 12 times, a rare privilege for someone born in the middle of the 20th century. Since the 1960s, aviation has separated travellers from the power and mystery of the oceans.

Like all visitors to England, my family marvelled at cathedrals and castles. Unlike most visitors of the 1950s and 1960s, we also marvelled at the cities that manufactured, and were manufactured by, the industrial revolution. My family lived in Manchester for a year, when coal smoke still fogged that city. We visited nearby Rochdale, where streams and canals riveted textile mills and whose weavers struck for cheaper grain imports in 1843. We climbed the seven hills of Sheffield, carved by the six rivers that were harnessed to power the wheels that sharpened the famous Sheffield steel.

Like many boys, I became fascinated by railroads and maps. I still have some of the maps of Britain I collected. One series compares Britain before and after the Industrial Revolution. Before the industrialization process set off on its dynamic, remorseless course, large green areas on the maps housed fewer than 12 inhabitants per square kilometre. In my mind, this was the world of castles, of the tales of King Arthur and Robin Hood, of forests and deer, of silk and leather.

Industrialization inevitably shrank these green and pleasant vistas, as networks of canals and railways, and clusters of factories and 'dark satanic mills' proliferated and altered the landscape. This was the world of novelist Charles Dickens and Oliver Twist, of brick and steam, of cities and coal. I found the industrial landscape fascinating – no less fantastic a stage for the imagination than undisturbed woodlands. And the industrial landscape reached from Britain around the world, conveyed in Kipling's 'great steamers white and gold', and Masefield's 'dirty British coasters'. Cotton came from plants sunning in Egypt and India, wool from sheep grazing in Australia, whole grain and flour from the wheat belt advancing westward across the US, while mundane, but vital, cargoes – timber from the Baltic, dairy produce from Denmark, fruit from the Mediterranean – came overseas from nearby Europe. Outwards from the workshop of the world went the money, the people and the 'Tyne coal, road-rail, pig-lead, firewood, iron-ware and cheap tin trays' that helped to spread the 'industrial spirit' across the globe.

The English landscape proved that the history of nature is, or became, the history of *human* nature. The History of Marine Animal Populations (HMAP) research programme of the Census of Marine Life (CoML) is demonstrating how the seascape, like the English landscape, has both influenced and been cultured by the courage, ingenuity and folly of human endeavour.

An orientation to the Census of Marine Life

During the late 1990s, several leading marine scientists shared their concerns with the Alfred P. Sloan Foundation that humanity's understanding of what lives in the oceans lagged far behind our desire and need to know. Some of the scientists emphasized the chance for exciting discoveries about the world in which we live. Much remains to be discovered about the diversity of life in the oceans. For example, ichthyologists have so far identified about 16,000 species of marine fishes. They also believe that about 4000 species of marine fishes remain to be discovered and described. The age of exploration in the oceans is not over. Other researchers highlighted the importance of establishing baseline information on the distribution of marine life. For most marine animals, we lack reliable maps of their range or distribution. Pointing to the changing abundance of many species, other researchers called for improved management of fisheries and marine reserves. They noted increasing exploitation of largely unsurveyed areas, such as the continental slopes and sea mounts, as well as violent debates about numbers of supposedly well-known species, such as cod, tuna and salmon.

Happily, these diverse scientists converged on a strategy to address their concerns: conduct a worldwide census whose purpose would be to assess and explain the diversity, distribution and abundance of marine life. The founders of the programme organized the Census of Marine Life (www.coml.org) around three grand questions:

1. What lived in the oceans?
2. What lives in the oceans?
3. What will live in the oceans?

Implicit in this outline research agenda is the recognition that a survey of contemporary marine life would have much more value if compared with historical information. Accordingly, the first grand question – what lived in the oceans? – motivated the HMAP programme that has created this book.

Commencing in 2000, CoML is a decade-long programme that will conclude with the First Census of Marine Life in 2010. While the 2000 or so researchers participating around the world are now three-quarters of the way through the ten-year period, the census will reach its highest peaks during the next couple of years in terms of effort and output, since all the components of the census work towards the 2010 synthesis. Alongside HMAP there are three other key elements of CoML:

1. It has 14 field projects, which are all making new observations in diverse realms to reveal much about 'what lives in the oceans'.
2. The question of 'what will live in the oceans' is addressed by the Future of Marine Animal Populations (FMAP) programme.
3. CoML's data assimilation framework, the Ocean Biogeographic Information System (OBIS) provides both archiving and access with regard to research results.

Opportunities for synergy between HMAP and the rest of CoML abound. HMAP researchers are already integrated with some of the ocean realm field projects – for example, in the Gulf of Maine area – and I wonder whether they might further enhance their cooperation with scientists surveying today to present a seamless past and present for that exemplary region, which has been exploited for almost 400 years. Both of the polar regions provide scope for integrating past and present. I wonder whether HMAP's walrus and seabird experts might link with CoML's Arctic team, and with researchers engaged in the circum-Antarctic Census of Marine Life, to document not just two or three but many decades of populations of albatross, of which 19 of 21 species are now endangered. Mention of walruses and birds leads to the global question of whether HMAP experts on top predators, including whales, seals and turtles, might join with the Tagging of Pacific Predators (TOPP) field project to create knowledge that spans long periods and wide oceans for these long-lived and far-ranging animals.

Links already exist between HMAP and FMAP. The former's millennial studies of the Wadden Sea have adapted analytic tools of the kind that ecologists have developed for prediction, while FMAP researchers have exploited more than 50 years of data on catches of long-line fishermen to learn and forecast trends in the diversity, distribution and abundance of tuna and billfish. Forecasting and back casting are two sides of the same coin, and researchers in HMAP and FMAP are sure to find mutual benefit by sharing data and tools.

HMAP is already a major provider of data to OBIS, which in 2007 had already facilitated access to about 13 million records of 80,000 species, and aims to provide tens of millions of records for more than 200,000 species. The extrapolation and digitization of historical data uncovered in archives located across the globe will be one of HMAP's great legacies. The integration of this material within larger databases will allow us to perceive patterns over larger areas, longer eras and covering more forms of life. OBIS interacts with historians by presenting the fruits of their research in new, innovative ways, and also by adapting its framework to accommodate data that were generated for various purposes, in many different forms, by an array of agencies that functioned long ago. Drawing on OBIS, everyone involved in the census can devise enthralling new visualizations. In an era when all information is born digital, CoML has the chance not only to perform analyses and create insights in words and numbers, but to share them in both static and dynamic images. Long appreciative of the power of maps, historians have much to contribute in making CoML innovative in the forms with which we share what we learn.

LIMITS TO KNOWLEDGE

I strongly believe that it helps to understand what is known and why we know what we know, what we do not know but might readily learn, and what it is very hard to learn or might be unknowable. In other words, it helps to understand the limits of knowledge. In the census we often refer to these limits by speaking of the known, the unknown and the unknowable. For both producers and consumers of knowledge, knowing what you do not know, like writing '*terra incognita*' on a map, can be as valuable as adding detail to lands that you have explored. Disclosing the limits to knowledge can also be among the most useful of acts. Such disclosure helps people to choose where to explore, avoid frustration and hedge bets.

The limits that separate knowledge into the known, unknown and unknowable are numerous and diverse. Like biodiversity, these limits call for taxonomy. Let me offer a taxonomy of limits to knowledge divided into five families:

1   The vastness of the present state.
2   Reasoning from the parts to the whole.
3   Surprise interventions from outside.
4   Blinders we put on ourselves.
5   The invisibility of the lost past.

The first family of limits, the vastness of the present state, encompasses the physical and practical barriers that are integral to ocean space. It is very hard to see what is far or dark or deep or at high pressure. A fundamental reason that knowledge of marine life remains crude is the imperviousness of water to light. The expanse of the ocean also challenges the timeliness and frequency of observations. Only a few governments regularly send out research vessels to make direct scientific measurements for assessments of stocks. While these ships use trawl nets with constraints upon their mesh size, they also trawl in only a few locations. Their sonar probes only narrow swathes that fish might avoid because of vessel noise or pressure waves. The survey vessels sample a tiny fraction of the sea, mostly near shore and at shallow depths. Technology offers stunning progress in observing marine life; but an ocean observing system that regularly reports a quite complete picture to us is many decades away.

Reasoning from the parts to the whole (the second family of limits) encompasses both statistical challenges and models. There are, and always have been, serious problems associated with fisheries statistics. Fishers tend to under-report catches, and commercial activity addresses *fished* stocks, rather than *fish* stocks. 'By-catch' data are used to fill in estimates of some other species, but also suffer biases. In turn, little is known about the validity of the mathematical models used to transform available data into assessments of stocks. For many species, the sparse knowledge of life cycles may limit the realism of models. Models themselves suffer limits of many kinds, including simplification, while mathematical forms are used, and errors occur in describing initial conditions. Few models capture extremely sensitive, non-linear systems where, for example, an ecological regime shifts from dominance of one species to another.

A tsunami exemplifies surprise interventions from outside, the third of the families. Shocks initiated by humans also belong to this group of limits. A prohibition against fishing would be a surprise intervention by people, as surprising to fish as the shock inflicted on Earth by a tsunami. Yet, as the history of human societies and natural history both testify, abrupt changes do, from time to time, disturb a seemingly orderly world.

The fourth limit (the blinders we put on ourselves) stems from economic as well as cultural factors. Limited funding is made available to the agencies that carry out stock surveys, such as the US National Oceanic and Atmospheric Administration, resulting in small or short samples. Governments in most countries have inadequate means to verify or rectify the numbers that they receive. Disciplinary myopia also causes experts to overlook data. Although microbial life might make up 90 per cent of ocean biomass by weight, researchers have largely ignored it until recently. We also all bring cultural biases to our work, which leads to the exclusion or discounting of certain data and information types. Often, we only know what someone else is willing to pay for.

A fifth family of limits, especially important for HMAP, is the invisibility of the lost past. Some phenomena leave no traces, or they may have left traces but we cannot find them. It is very difficult to reason backwards to what has disappeared. The tendency of successive generations of scientists to supplant the measuring points deployed by their predecessors with yardsticks of their own, the so-called 'shifting baseline syndrome', falls into this category of 'lost in the past'.

In essence, I urge researchers to continue confronting limits. By identifying them openly, we have a better chance of overcoming them, or to help other people act wisely when faced with hard and seemingly insurmountable limits to knowledge.

Hypothesis: The great reversal

Let me now return to the environment itself. During the 1980s, I studied long-term trends regarding the use of energy and materials. I learned that their efficiency has improved for as long as records are available. A steady increase in the sparing of these resources per unit of output or service provided is normal. In 1992, curious about agriculture, I asked my esteemed colleague, agronomist Paul Waggoner: 'How much land can 10 billion people spare for nature?' By 1996, I began to hypothesize that humanity had reached an inflection point in development about 1970, when the rate of human population growth peaked, allowing improving efficiency to reverse the expansion of cropland in several nations. Seeing expanding forests in several nations and borrowing a phrase from the demographers of *population transition*, geographers now speak of a *forest transition*. Examples from energy, materials, farming, forestry and water use, too, suggested that a 'Great Reversal' of resource use was occurring in several ways.

During the mid 20th century, humans began to reverse the pattern they had followed for millennia of extending further into nature to meet needs for food and materials. Recognizing this Great Reversal, I and my associates explored the areas in human use for cities, logging and farming. We searched for principles and trends to forecast land use in the late 21st century to accommodate a world of 10 billion people. Offsetting the sprawl of cities, rising yields in farms and forests, and changing tastes can release large amounts of land. For example, even with growing population and cities, the US could still newly spare for nature an area twice the size of Spain in the next century. Cutting of forests peaked in the US in 1906, and the volume of both hardwoods and softwoods standing in the nation's forests has risen steadily since about 1950. In fact, studies of forest biomass for the 1990s in the boreal and temperate region in more than 50 countries show the forests expanding in area and/or volume. Globally, a wise and intelligent humanity could extend the 'Great Reversal' into a 'Great Restoration' of nature on land.

What about in the sea? Is a reversal near, and a restoration in prospect, for fish as it may be for trees? Answering the questions of marine reversal and restoration requires answers to two questions: what *lives* in the oceans and what *lived* in the oceans? Here again we come to HMAP. I suspect that human use of the oceans is 100 years behind our use of the land – that is, the forest transition or Great Reversal in land use may precede and foreshadow a Great Reversal in the utilization of marine resources. The evidence of HMAP may reveal the long-term course of exploitation, including unwise exploitation, of the oceans. HMAP may give a head start of decades and even centuries in anticipating trends – improving as well as declining. Use of overarching hypotheses, such as the Great Reversal on the land, may help HMAP to navigate its work in the sea.

HMAP: Ingenious evidence and tasks

Speaking of evidence, I want to remark further on the ingenuity of HMAP researchers: congratulations on discovering new sources of evidence, and also on finding new evidence in old sources. Along with menus, HMAP's evidence includes buttons and bones, logbooks and lore, paintings and pavements, isotopes and ice. HMAP researchers keep extending the limits of knowledge by finding new ways of making the past visible. They help us to lift self-imposed blinders on what constitutes useful source material. But more information can still be extrapolated from traditional sources. I like to read early texts of geography and history, and since HMAP was launched I have looked at such sources with fresh, wide eyes. Let me briefly mention some indicative material that might be fruitful to revisit. Starting earliest, I think of Seneca's *Naturales Quaestiones* and his treatises on natural phenomena. Another source might be the chronicles of St Brendan, who, it is believed, was born in 484 AD near Tralee, County Kerry, Ireland, where he died in 577 AD. His *Navigatio*, or wandering, was assembled about 300 years later. Some is pure Celtic fantasy, but to me his description of one region as 'curdled ocean' sounds very much like it could be the Sargasso Sea. A fellow Irishman, Dicuil Hiberniae, wrote *De Mensura Orbis Terrae* (*On the Measure of the Earth's Globe*) in about 825 AD. It is important for providing the first records of Iceland and the Faeroe Islands, as well as the Orkneys and Shetlands. Maybe Dicuil's work is useful for HMAP.

The medieval Icelandic chronicles, the *Graenlendinga Saga* and also *Eirik's Saga* famously offer the primary written evidence for the Norse landfall in North America. We should look again at their accounts through the lens of HMAP. About the same time as Eirik, the Moroccan geographer and cartographer al-Idrisi wrote a Universal Geography, produced in Sicily in 1153. His *Nuzhat al-Mustaq* describes the seas of the North Atlantic: 'There are animals of such great size that the inhabitants of the islands use their bones and vertebrae in place of wood to build houses. They make hammers, arrows, spears, knives, seats, steps, and in general every sort of thing elsewhere made of wood.' Perhaps HMAP can extract some ideas from al-Idrisi about the distribution and size of Atlantic marine animals 950 years ago.

I hope such fantastical ideas stimulate the imaginations of historians and ecologists. The point is, by surmounting limits to knowledge, we can construct the history of marine animal populations. This brings me to what I see as the three essential tasks of HMAP. The first is to create pictures of what lived in the oceans before fishing became important. In some places, the time is 50 years ago,

500 in others and 1000 or more in a few. More broadly, we might compose a set of snapshots of marine life, say, in year 1, year 1000, and years 1500, 1600, 1700, 1800, 1850, 1900, 1925 and 1950. Second, HMAP is required to dissect the influence of fishing, climate variability and other factors driving changes in marine animal populations since fishing became important. Nature includes human nature and so must its history. The third task is to create and make accessible long time series on marine animal populations and related factors so that future researchers can study the processes of change in the marine environment more effectively. I hope our legacy of data will show not only a course of exploitation, but the path to a Great Reversal and Great Restoration of marine life.

The work undertaken since 2000 suggests that the HMAP community is growing strong. Its field of enquiry, in which historical marine ecology and maritime history are integrated, is as valid within history as the study of the industrialization of Britain. In fact, it is fundamentally the same. As the exciting research presented in this volume shows, by recognizing and lowering limits to knowledge, we can write an epic history: global, deep and human, with imagination and with profit for nature. *Oceans Past* shows the way forward. By continuing on this path down to the First Census of Marine Life in 2010, HMAP will compile richly documented and beautiful answers to the question: what did live in the oceans?

Acknowledgements

Thanks to Poul Holm, Andrew Rosenberg, Tim D. Smith, David J. Starkey and Paul Waggoner.

1

Oceans Past:
History Meets Marine Science

David J. Starkey, Poul Holm and Michaela Barnard

The chapters that comprise this volume were originally presented as papers to an international conference – Oceans Past: Multidisciplinary Perspectives on the History of Marine Animal Populations – held in Kolding, Denmark, in October 2005. This meeting was convened by the History of Marine Animal Populations (HMAP) project, which forms the historical component of a global network of researchers engaged in the Census of Marine Life (CoML), a ten-year initiative designed to assess and explain the diversity, distribution and abundance of marine life in the oceans – past, present and future (see the Foreword). The principal aim of the conference was to bring together researchers from the natural sciences and the humanities to discuss how and why marine life in the world's oceans has changed over time, and the role that humans have played in that dynamic process. Such an ambitious remit permitted those engaged in HMAP to take stock of the headway their project had made since its launch in 2001, as well as sharing their findings with the scientific community at large.

Ten of the 75 papers delivered to the conference have been revised for presentation in the pages that follow (a further selection has been published in MacKenzie and Ojaveer, 2007). They reflect the breadth and depth of the HMAP project in various ways. In personnel terms, there are contributions from researchers based in Russia, Canada, Australia, Denmark and the UK, although a clear majority of the authors hails from the US. Their disciplinary mix is more pronounced, as befits a conference that attracted archaeologists, oceanographers, biologists, ecologists, historians and social scientists to debate the character and significance of the long-term interaction of human societies and the marine environment. It is notable, moreover, that in a work which examines the past to shed light on the present, only a few of the authors and editors are historians by profession.

In temporal terms, the contributions to this volume focus on the 19th and 20th centuries. Their topical scope, however, is much broader, with coverage of species ranging – as the order of contents indicates – from common periwinkle snails to sponges, crustaceans, wildfowl, alewives, cod, whales and human scientists and policy-makers. The spatial parameter is extensive, too, for as well as embracing the open oceans inhabited by whales and whalers, the chapters consider the inshore waters and continental shelves populated by a myriad of marine animals, together with the conference halls where human politicians, administrators and stakeholders devise environmental management regimes.

Amid these different spectra, a singular message emerges from this volume: that knowledge and understanding of oceans past are critical to managing marine resources in the present and in the future. This is not to suggest that history alone can explain the current status of the seas and oceans. Rather, it should be taken to mean that reducing the constraints on knowledge imposed by the 'lost past' is essential to a holistic understanding of the complex issues inherent to the interaction of human societies and the marine environment (see the Foreword and Afterword). Retrieving and analysing fragments of the past instils a dynamic quality into the equilibrium-based theories that have traditionally marked the approach of marine scientists (Holm et al, 2001, ppxiii–xix). This can be achieved by measuring change over time, either in quantitative terms through the calculation of numbers of fish, fishers and species in particular localities at specific points in time, or in a qualitative sense through the memories and memorials of those who observed 'things as they were then'. Measurements and memories, in turn, facilitate analysis of causality and assessments of significance; history, in other words, can help answer two questions – why? so what? – that are often beyond the reach of the natural sciences.

Underpinning the contribution of history to this holistic understanding is evidence of the past. This is invariably and exceptionally partial: partial in so far as it is a remnant of the information that once existed, and partial in that it reflects the perspective, or bias, of its originator. To make a meaningful contribution, history therefore has to submit its exceptionally partial data to a rigorous process of validation in order to ascertain the extent, meaning and shortcomings of the information. This entails undertaking a provenance of the primary source material designed to establish why it was generated, by whom, by what methods, and how and why it was preserved. Corroboration is also part of this process, for if the provenance tests positively against information derived from another source, then confidence in the reliability of the data is enhanced. Assessing the deployment of similar data by previous researchers is a further corroborative device that can be used to gauge the accuracy of the material.

Armed with evidence validated in this way, history has the capacity to address issues that cross disciplinary boundaries. In the present study, history meets marine science and thereby adds significantly to knowledge of the marine environment by locating, validating and applying new evidence pertaining to fish stocks, fishing effort and/or the relationship between these two variables. In this respect, value is added by pushing back in time the frontiers of what is known. History can also operate 'laterally' by generating knowledge of particular cases that confirm, refute or modify the findings of existing case studies. More importantly, perhaps, history possesses an interpretative quality that enhances understanding of the causes and ramifications of environmental change. This contribution assumes one of three guises. The first is context, an understanding of the setting in which an event or activity takes place, and of the place of that event or activity in that setting. Second, an empathy with how people live(d) in different temporal and geographical locations is a form of understanding that is vital to historical scholarship. Context and empathy are integral to the third guise, for history, *inter alia*, is a social science that seeks to understand why individuals and societies decided – or were compelled by anthropogenic factors and/or natural forces – to behave as they did at particular points in time.

While evidence is critical to history, this volume offers evidence of the significance of history to knowledge and a holistic understanding of the patterns and processes of change in the marine environment. It demonstrates, for instance, the value of validated primary information. All of the authors base their analyses on data gleaned from primary sources of various provenance. Among those utilized are archaeological finds, newspaper reports, business advertisements, municipal archives, fishing logbooks, abstracts of whaling logbooks, trade journals, personal records and state departmental memoranda. While these sources have been assessed by each respective author, an appraisal of the quality of the evidence yielded is a prominent theme in a number of contributions. Stefan Claesson, for example, discusses the potential of the extraordinary detail provided in the logbooks submitted by New England fishermen during the 1850s. In a similar vein, Julia Lajus assesses the archival, as well as the intellectual, legacies of Karl Ernst von Baer, while Tim D. Smith, Elizabeth A. Josephson, John L. Bannister and their colleagues test the veracity of the evidence derived from logbooks by Maury, Townsend and other 19th- and 20th-century researchers.

Analysis of these various source materials offers numerous insights into the management of marine resources. Glenn A. Jones shows that the retail price of food can be tracked back to the 1850s by processing information contained in American restaurant menus – that is, he provides evidence that predates by a century or so the first equivalent 'official' price data. Likewise, Stefan Claesson describes how the raw data contained in contemporary logbooks have been utilized to estimate the stock of cod that inhabited the Scotian Shelf during the 1850s, a baseline that facilitates comparison with stock assessments undertaken in the same area in the early 21st century (Rosenberg et al, 2005). The breadth, as well as the temporal reach, of what is known is also improved in this volume. April M. H. Blakeslee's study of the spread of the common periwinkle snail in northeastern North America provides further evidence of the role of human agents in the invasion of marine ecosystems by species from distant habitats. Loren McClenachan's investigation into sponge fishing in Florida's inshore waters is similarly instructive in its revelation that human harvesters not only depleted the sponge banks, but also created conditions that fostered the spread of the disease which ultimately proved fatal to this once-rich marine resource.

Understanding of the rationale of past marine resource utilization is enhanced by this volume. Julia Lajus's chapter implies that Karl Ernst von Baer, in extolling the virtues of history as an explanatory tool, was very much a 'fish out of water' in the context of the scientific community of the mid 19th century. Context is the key theme of Carmel Finley's contribution, for it was the geo-political objectives of the US during the 1950s that demanded the implementation of maximum sustained yield; in essence, the exigencies of the Cold War fashioned a policy that had a minimal scientific base, but a major and largely negative impact on global fish stocks. Matthew McKenzie also considers the development of world systems in his examination of the alewife runs of New England. Against this background, he narrows his focus to empathize with the people of the Cape Cod settlements who had long appreciated the importance of stocks of river herring to their way of living before the 1860s. In improving our understanding of why this communal interest changed, he explains how, over a period of three decades, this species had all but disappeared from the rivers that coursed through these growing townships.

Societal, communal and individual behaviour is a theme that is central to history. The decisions of states, institutions and people to fish particular species more or less intensively have long-term impacts on marine animal populations and the environments that they inhabit. In turn, this influences the welfare of societies, communities and individuals, as well as the choices available to future generations. Such a dynamic, interactive relationship imbues this volume; indeed, it emerges whenever history meets marine science. But it is brought into sharp relief by Glenn A. Jones's study of seafood consumption patterns in the US since circa 1850. As demand for a naturally occurring commodity increased, so production intensified and the supply of that foodstuff became scarcer. As a consequence, the price increased consistently and dramatically, and consumption fell. Whereas it was not until late in this cycle that the harvesting of marine animals was restricted, no limit – other than that arising from the 'discipline' of market forces – was imposed on human consumption.

This is merely one of many insights from oceans past that historians, scientists and those charged with managing ocean resources will find in this volume.

References

Holm, P., Smith, T. D. and Starkey, D. J. (eds) (2001) *The Exploited Seas: New Directions for Marine Environmental History*, International Maritime Economic History Association, St John's, Newfoundland, Canada

MacKenzie, B. R. and Ojaveer, H. O. (eds) (2007) 'History of marine animal populations and their exploitation in northern Europe', *Fisheries Research*, Special Issue, vol 87, nos 2–3

Rosenberg, A. A., Bolster, W. J., Alexander, K. E., Leavenworth, W. B., Cooper, A. B. and McKenzie, M. G. (2005) 'The history of ocean resources: Modelling cod biomass using historical records', *Frontiers in Ecology and the Environment*, vol 3, pp84–90

2

Invasive or Native? The Case History of the Common Periwinkle Snail (*Littorina littorea*) in Northeast North America

April M. H. Blakeslee

In recent years, species invasions have become recognized as one of the major human-mediated impacts on natural systems. Specifically, invasions in marine environments have caused considerable harm to native communities, as exemplified by San Francisco Bay on the United States (US) west coast. From 1961 to 1995, with an average of one new species invading every 14 weeks, these marine exotics comprised 40 to 100 per cent of the common species within the bay, up to 97 per cent of its species abundance and up to 99 per cent of the species biomass. Due to this high rate of invasion, the bay has become one of the world's most heavily invaded aquatic ecosystems, a factor that has greatly contributed to its high level of degradation (Cohen and Carlton, 1998). Invasions can therefore have profound effects on ecosystems, with significant implications for ecological and management work.

One major dilemma in the study of invasion biology is that not all species observed in a particular ecosystem can be demonstrably classified as native or non-native. These species are referred to as cryptogenic and can comprise a considerable proportion of species observed in a particular region (Carlton, 1996). For example, Carlton (1996) found that possible invasions – i.e. cryptogenic species – constituted as many as one third of the number of known invasions in San Francisco Bay. The number of cryptogenic species that prove to be non-indigenous on the east coast of the US will probably be higher than on the west coast because Europeans settled there much earlier and long before rigorous inventories of native biota were consistently undertaken (Ruiz et al, 2000). Almost 30 per cent of the species richness found in Chesapeake Bay, for example, may actually be cryptogenic (Ruiz et al, 2000). For resource managers and conservation biologists charged with restoring and maintaining native species, biological diversity and ecosystem function and services, knowledge of the native members of the community is a fundamental starting point, especially as little can be done (or is likely to be done) to address the potential impacts of these species when it is uncertain whether or not they are non-native. The resolution of the ambiguous ecological history of cryptogenic species is a critical conservation goal, particularly with regard to conspicuous and abundant species. Even if the removal of a given species may now be impossible, other conservation and management goals may then justifiably take priority, such as protecting truly native or endangered species.

When the evidence for invasions is not obvious (typically because they occurred prior to human exploration of the oceans and their coastlines), the application of 'criteria', or criteria-based evidence for introductions, can be deployed to resolve their native or non-native status (e.g. Chapman and Carlton, 1991). Such criteria must be based on evidence derived from multiple sources, including geographical data, the biology and ecology of the species, historical material and genetics (Chapman et al, 2007; Chapman et al, in press). Close associations with human transport mechanisms and food supplies, for instance, often infer that species were introduced by anthropogenic agents (Chapman et al, 2007). Likewise, the absence of a fossil record in the purported founding population, together with the lack of a natural dispersal mechanism, such as 'stepping stone' invasion, are also potential signatures for an introduction, while strong genetic bottlenecks can result from recent founding events (Grosberg and Cunningham, 2000). Because these criteria indicate whether a species is native or non-native, they can be used as signatures for resolving cryptogenic histories.

One North American cryptogenic species, whose status as native or non-native has been debated for the past 150 years, is the marine snail, or common periwinkle (*Littorina littorea*) (see, for example, Ganong, 1886; Clarke and Erskine, 1961; Clarke, 1963; Berger, 1977; Wares et al, 2002; Chapman et al, 2007; Chapman et al, in press; Wares and Blakeslee, 2007). A model species for marine introductions, this snail has profoundly affected marine communities, making the resolution of its cryptogenic status even more important. Furthermore, because the cryptogenic status of a species often stems from incomplete historical knowledge of its presence in a region (Ruiz et al, 2000), the lingering ambiguity surrounding *Littorina littorea* is particularly surprising given the significant volume of historical, archaeological, ecological and genetic data that have been assembled on this species over several decades (e.g. Ganong, 1886; Clarke and Erskine, 1961; Clarke, 1963; Bird, 1968; Berger, 1977; Carlton, 1982; Wares et al, 2002; Chapman et al, 2007; Wares and Blakeslee, 2007). This chapter reviews the evidence pertaining to *L. littorea*'s range, spread and impacts on North American populations. In so doing, it adds depth and insight to the literature that has been generated through the long running efforts of scientists to resolve the snail's North American cryptogenic history.

LITTORINA LITTOREA: ECOLOGY, HISTORICAL DISTRIBUTION AND IMPACTS

Littorina littorea is a marine gastropod with planktotrophic larvae that exists in widely varying environmental conditions, including the rocky intertidal zone, estuarine and, at times commonly, both sandy and muddy environments (Moore, 1937; Brenchley and Carlton, 1983; Reid, 1996). Snails can attain relatively large shell lengths, the largest individual ever recorded being 52.8mm (Reid, 1996), while the largest individual personally observed was 33.71mm. Furthermore, populations of *L. littorea* can be extremely abundant, reaching numbers upwards of 150 individuals per 0.05 square metre (personal observation), which converts to about 3000 individuals per square metre.

Littorina littorea is currently found in two large North Atlantic populations, Europe and North America, but is absent from the Faeroes, Iceland, Greenland and other mid-Atlantic islands. In Europe, *L. littorea* is found from the White Sea in northern Russia to Portugal (Reid, 1996), whereas in North America established populations are evident from Red Bay, Labrador, to Cape May, New Jersey (Reid, 1996; personal observation), though ephemeral populations have been found as far south as Virginia (Reid, 1996). *L. littorea* is known to be native to Europe, but its presence in North America is less well understood. What is known about the presence of *Littorina littorea* in North America is that during the 1850s, it spread rapidly and sequentially southwards from Halifax, Nova Scotia (with oral reports placing it about 15 years earlier in Pictou, Nova Scotia), into the US, reaching Cape May just 30 years later. This invasion was well documented in both primary and secondary sources and its pattern pieced together by Carlton (1982) and Steneck and Carlton (2001).

The influence of this spread on intertidal communities is also well known. *Littorina littorea* is the dominant herbivore in the US rocky intertidal zone and has greatly impacted upon, and altered, algal communities (Lubchenco and Menge, 1978; Lubchenco, 1983; Bertness, 1984), as well as competitively displacing native snails, including *L. saxatilis* (rough periwinkle snail) and *Ilyanassa obsoleta* (eastern mud snail), in areas where it is abundant (Yamada and Mansour, 1987; Brenchley and Carlton, 1983).

THE CRYPTOGENIC STATUS OF *LITTORINA LITTOREA*

Although the invasion of *Littorina littorea* into the US, and its impacts upon marine flora and fauna in the intertidal zone, have been well documented over the last 150 years, uncertainty regarding its status as native or non-native in North America has persisted to this day. What remains unclear is whether the Canadian population from which the spread into the US originated was native and confined to the Maritime provinces until the 1850s, or whether it was an anthropogenically introduced population from Europe that was first established in Canada and then invaded southwards in the mid 19th century (Reid, 1996). Due to this uncertainty, *L. littorea* is currently considered cryptogenic in North America.

The puzzle over the ecological history of *Littorina littorea* in northeast North America has been assessed by several researchers since 1886. Their investigations have generated historical, archaeological, ecological and genetic evidence (see Table 2.1), the validity of which is appraised in the sections that follow.

Historical evidence

The first sighting of *Littorina littorea* occurred at Pictou, Nova Scotia, during the early 1840s. This was reported retrospectively by J. W. Dawson, who claimed that he had collected individuals approximately 30 years prior to Verrill's 1874 account (reported by Verrill, 1874). In their appraisals of this 'discovery', many scientists believed that the snail had been recently and accidentally introduced from Europe. Yet others (including Dr Dawson) were convinced that it was native to North America and had been overlooked by conchologists (Verrill, 1874). Willis (1863) stated that several older Nova Scotian inhabitants had testified that they had collected periwinkles as 'school boys', which he interpreted as implying that the periwinkle was indigenous to North America. Ganong (1886), however, considered this claim to be unscientific and suggested that *L. littorea* might have been confused with native periwinkles (Chapman et al, 2007).

While the oldest records of *L. littorea* in North America derive from sightings and oral testimonies, the first live specimens were collected at Halifax, Nova Scotia, in 1854 and are now preserved at the Smithsonian Institution (Willis, 1863; Ganong, 1886; Chapman et al, 2007). Willis stated that several eminent British conchologists were consulted when these specimens were found, all of whom were surprised that the common periwinkle snail existed in North America. Until about 1870, *L. littorea* was unknown to American conchologists (Bequaert, 1943), while in dozens of mid 19th-century shell catalogues, the presence of rocky intertidal and estuarine snails such as *L. saxatilis*, *L. obtusata*, and *Ilyanassa obsoleta* throughout northeast North America was recorded (e.g. Binney, 1863; Gould, 1851), but *L. littorea* was never listed. Steneck and Carlton (2001) doubted that such a conspicuous species, especially one commonly utilized as a food source in Europe, could have been overlooked by 19th-century naturalists (Packard, 1870; Reid, 1996).

The introduction of *L. littorea* from Europe may have occurred intentionally through its import as a food source, or accidentally as a result of conveyance with ballast rocks (Reid, 1996; Steneck and Carlton, 2001). Several authors have supported the former explanation, notably Packard (1870), who suggested that 'this species is identical with the common periwinkle of the English coast and its increase

Table 2.1 *Evidence relating to the cryptogenic status of* Littorina littorea *in northeast North America*

	Native to North America	
Type of data:	*Evidence for:*	*Citation:*
Archaeological evidence	Pre-Columbian shells found at Nova Scotian and Newfoundland sites in Maritime Canada.	Clarke and Erskine, 1961; Clarke, 1963; Bird, 1968; Wagner, 1977
Ecological/natural history evidence	Unfavourable currents preventing dispersal southwards from Canadian Maritime locations along the Northumberland Strait but enhanced shipping between Canada and US in mid 1800s may have allowed spread into US.	Clarke and Erskine, 1961
Molecular evidence	Fixed differences in allozyme data. Unique haplotypes in mtDNA data.	Berger, 1977 Wares et al, 2002
	Introduced to North America	
Type of data:	*Evidence for:*	*Citation:*
Historical evidence	Rapid and sequential spread from a single point of origin.	Steneck and Carlton, 2001
	Early American conchologists had no record of *Littorina littorea* prior to 1870.	Bequaert, 1943
	European conchologists expressed surprise at *Littorina littorea*'s presence in North America.	Willis, 1863; Verrill, 1874
	Suggestion that intentional introduction as a food source would have been welcomed.	Packard, 1870
	Evidence of rock ballast mediated introductions in other species.	Coyer et al, 2006
Archaeological evidence	No fossil record in the US. Only a handful of specimens found in Maritime Canada; those that were radiocarbon dated do not predate Norse visits to Maritime Canada.	Chapman et al, 2007
Ecological/natural history evidence	*Littorina littorea* is not found in Iceland or Greenland, believed to be stepping stones for many North American rocky intertidal species.	Ganong, 1886; Johannesson, 1988; Ingolfsson, 1992; Reid, 1996
	Littorina littorea larvae cannot survive direct crossing from Europe to North America.	Kraeuter, 1976
Molecular evidence	No fixed differences in allozyme data. No fixed differences in AFLP data.	Morris, 1979 Wares and Blakeslee, 2007

Note: The table lists the known historical, archaeological, ecological and molecular evidence for (a) a native origin for *L. littorea* into North America; and (b) an introduction of *L. littorea* into North America from Europe.

may be hoped for as it will introduce a new article of food to our poorer classes'. Furthermore, Steneck and Carlton (2001) claimed an intentional introduction of *L. littorea* by Europeans was a likely explanation for its North American presence, while Spjeldnaes and Henningsmoen (1963a; 1963b) suggested that the Vikings may have deliberately carried *L. littorea* snails across the North Atlantic to augment their diets. Alternatively, unintentional introduction via ballast rocks may have been the cause of the occurrence of *L. littorea* – and other species, like the seaweed *Fucus serratus* (Coyer et al., 2006) – in North America. Intact ballast rocks, probably covered in local flora and fauna, were often collected directly from the intertidal zone near European ports and then deposited at arrival ports in North America, a vector through which several European species may have been introduced to North America (Lindroth, 1957; Spjeldnaes and Henningsmoen, 1963a, b; Carlton, 1982; Chapman et al, 2007).

Archaeological evidence

Many more *Littorina littorea* specimens have been found in Europe than in North America (Reid, 1996). No verified fossil record exists for *L. littorea* in the US (Steneck and Carlton, 2001), although 19 specimens have been found in archaeological sites in maritime Canada (Chapman et al, 2007). One of these specimens collected from Nova Scotia was dated at 40,000 years before present (ybp) (Wagner, 1977), but the dating was based on stratification and not radiocarbon dating; thus its accuracy has been questioned (Chapman et al, 2007). The majority of the remaining archaeological specimens have been found at Nova Scotian sites and have been dated at ~500 to ~1000 ybp using radiocarbon dating (Clarke and Erskine, 1961; Clarke, 1963). Two additional 'ancient' specimens were found at a Newfoundland site (Bird, 1968), but these were not radiocarbon dated (Chapman et al, 2007). On the whole, the shells dated using radiocarbon methods have not predated Norse visits to the maritime region.

Ecological evidence

If *Littorina littorea* was native to North America, why did it suddenly expand southwards in the mid 19th century and not many years before? This is a puzzling question for those advocating a native North American origin for the snail (Reid, 1996). Clarke and Erskine (1961) suggested that unfavourable currents may have prevented *L. littorea* from dispersing from the Northumberland Strait to the coast of Nova Scotia and that it only spread southwards when commercial shipping became more prevalent. However, trans-Atlantic shipping during this period was also more prevalent; thus the explanation of enhanced commercial shipping does not limit the snail population that invaded the US to an exclusively Canadian origin. Moreover, unfavourable currents did not prevent the spread southwards of marine rocky intertidal species such as *Semibalanus balanoides* (acorn barnacle) and *Mytilus edulis* (blue mussel), which exhibit similar dispersal mechanisms to *L. littorea* (Wares and Cunningham, 2001). At the same time, the notion propounded by Wares et al (2002) that ecological change allowed the snail to be released from its southern Canadian confinement is corroborated neither by historical evidence nor by the experience of other species that have exhibited similar range expansions (Reid, 1996; J. Carlton, pers comm).

A direct crossing from Europe to North America by larval *Littorina littorea* is also thought to be highly unlikely (Reid, 1996), if not 'impossible' (Kraeuter, 1976), due to the survivability of the larvae in relation to the distance travelled (approximately 3000 km) (Reid, 1996). The case for a 'stepping stone' invasion via the Faeroes, Iceland and Greenland – which may have been the natural means by which *L. saxatilis* and *L. obtusata* expanded to the New World – has also been rejected since no trace of *L. littorea* has been found on any of these North Atlantic islands (Ganong, 1886; Johannesson, 1988; Reid 1996). Although *L. littorea* has pelagic larvae (unlike its direct-developing congeners), such broadcast spawning species often have trouble retaining and establishing populations in small areas, especially islands (Johannesson, 1988; Byers and Pringle, 2006). One further means of natural movement from Europe to North America – rafting on driftwood – has also been discussed (Bird, 1968); however, drifting *L. littorea* adults would have difficulty establishing themselves on stepping stone islands (Johannesson, 1988), while a direct crossing of drifting adults from Europe to North America would probably end up south of most of the current North American range of *L. littorea* and certainly far to the south of its first reported sighting in Nova Scotia (Kraeuter, 1976).

Molecular genetics evidence

A variety of molecular techniques has been used to ascertain the origins of the North American *Littorina littorea*, from allozyme analysis in the mid 1970s to DNA sequencing and amplified fragment length polymorphisms (AFLPs) during more recent times. The first of these molecular studies was performed in 1977. Berger (1977) used protein electrophoresis to compare allozyme (= enzyme) variability in *L. littorea* populations in Cape Cod, the US, and Roscoff, France. He determined that the allozyme diversity in the Roscoff population was much greater than in the Cape Cod population and that the number of alleles shared between the two populations was small. He suggested that fixed differences found at 7 out of 12 allozyme loci suggested an ancient population divergence and that, as a result, *L. littorea* was native to North America. However, these conclusions were questioned by Reid (1996) on the grounds that a similar investigation by Morris (1979) yielded a very different result. Morris compared the allozyme variability in *L. littorea* populations in Maine to those in Wales and found that the allelic composition of the two populations differed significantly, indicating that *L. littorea* was introduced recently to North America. Reviewing these opposing studies, Johannesson (1992) expressed surprise at the large differences in the allozyme patterns between the Cape Cod and Roscoff samples presented by Berger (1977). She noted that several allozyme studies conducted at a range of European sites reported much lower overall variability in alleles (= variant forms of a gene), and concluded that the Roscoff population was probably one of a few localities that possessed an unusually high number of alleles. Berger's conclusions have also been discredited by three other studies. Whereas Chapman et al (2007) noted that Berger's Roscoff sample was too limited geographically to reveal the range of genetic diversity within the European population, Cunningham (in press) compared the Berger and Morris datasets and concluded that the similarities between allele frequencies found at the North American site sampled by both, and the Welsh population sampled by Morris, suggested that 'something was amiss in Berger's allozyme study from France'. A third investigation, undertaken by Wares and Blakeslee (2007), generated an AFLP dataset (which explores numerous loci throughout the entire genome) and revealed zero fixed differences between the North American and European populations out of 120 scored AFLP loci. This finding was 'in stark contrast' to the conclusions yielded by Berger's smaller and less representative dataset.

Adopting a different approach, Wares et al (2002) used mitochondrial and nuclear DNA sequencing and reported 'unique' American haplotypes (= different forms of a mitochondrial gene) that were not observed in their European samples. Using Nei and Li's (1979) measure of net pair-wise divergence and the MDIV computer program (Nielsen and Wakeley, 2001), they found divergence estimates between Europe and North America that were at least 8000 years ago, suggesting that *L. littorea* was indigenous to North America. However, the study's total snail (North America: n = 57; Europe: n = 60) and site replication (North America: n = 5; Europe: n = 4) were small and only included specimens from a portion of the entire range of *L. littorea*. In addition, its mitochondrial data were suggestive of a severe genetic bottleneck in North America due to the significantly lower number of haplotypes (10) compared to the total evident in Europe (32). North American diversity made up only 24 per cent of the total diversity of the two populations. Moreover, native populations generally exhibit high allelic diversity and many rare alleles (and few common ones) due to the complex genetic structure arising from their long evolutionary history. In contrast, as the result of a genetic bottleneck, introduced populations are expected to show lower genetic diversity and contain few rare, and many common, alleles because common founding alleles will attain high frequency as the population expands, while rare alleles are often lost due to genetic drift (e.g. Ledig et al, 1999). Wares et al (2002, Appendix) had significantly fewer rare haplotypes in North America compared to Europe, a pattern that is expected in founding populations (Chapman et al, 2007).

Wares et al (2002) also had a small number of sequences (n = 18; 8 North American and 10 European individuals) from the nuclear ribosomal internal transcribed region (ITS) and found no shared alleles between European and North American ITS sequences. This, they claimed, also suggested a native origin for *L. littorea* in North America. However, because confidence intervals for divergence estimates using the ITS data did not exclude zero, this evidence is not conclusive. Additionally, Wares and Blakeslee (2007) suggested that the Wares et al's (2002) ITS data may have been inappropriate with regard to the origins of *L. littorea* because ITS data may harbour significant diversity for a number of reasons, including high copy number, high substitution rate and low selective constraint. Accordingly, the sample size used by Wares et al (2002) was too small to discern levels of diversity within, and between, the two populations and was therefore inadequate as a means of determining ancient population divergence.

One of Wares et al's (2002) main contentions was that unique haplotypes observed in North America were not apparent in Europe, suggesting population divergence in the distant past. This is because in a recent introduction all the alleles found at a locus in the founding population should have been derived from the source population (Grosberg and Cunningham, 2000). However, two recent investigations have shown that the study's small sample size (especially in Europe) probably precluded the discovery of European haplotypes that corresponded to the unique ones observed across the Atlantic. First, Wares and Blakeslee (2007) demonstrated this point graphically (see Figure 2.1). Visibly asymptoting accumulation curves are indicative of the total diversity in a population (Gotelli and Colwell, 2001). Figure 2.1 clearly shows that the North American curve is asymptoting, while the European curve is linear (R^2 = 0.994). This result infers that continued sampling in Europe would reveal more haplotypes, while the discovery of haplotypes in North America appears to be levelling off. Overall, this evidence demonstrates the difficulty in satisfying one of the requirements for an introduction – that all the alleles found at a locus in the recipient biota should be derived from the source population – when the source population has so much genetic structure compared to the recipient population (i.e. it is easy to 'miss' alleles in a highly diverse source population).

The second investigation (Chapman et al, 2007) used a probability analysis to reveal that the number of 'unique' haplotypes witnessed by Wares et al (2002) in North America fell within the number of unshared haplotypes that would be expected based on the number of unique haplotypes found in Europe. This strongly suggests that these 'unique' haplotypes may, in fact, exist in Europe, but due to insufficient sampling were not found. In fact, based on the number of rare (= 1 occurrence) haplotypes observed in Europe in Wares et al's (2002) dataset, the genetic diversity in Europe should be much higher than in North America. In the analysis shown in Figure 2.2, Europe is predicted to have 38 times more haplotypes than North America, which again points to the inadequacy of sampling in the native population. In fact, other investigations of known introductions have made similar observations, i.e. the discovery of 'unique' haplotypes in the non-native population. For example, Miura et al (2006) found that a trematode species

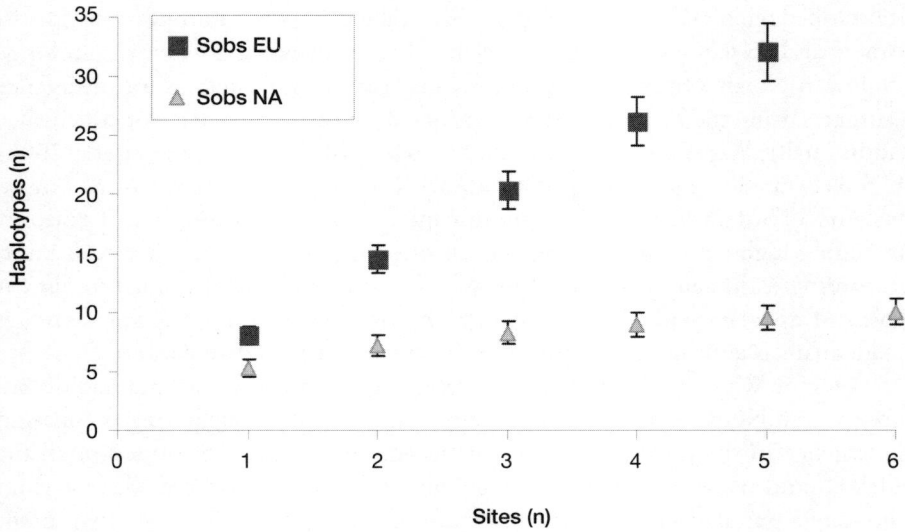

Figure 2.1 *Rarefaction curves for haplotype diversity and sampling effort*

Notes: This figure portrays the accumulation curves (S_{obs}) for the number (n) of European (EU) and North American (NA) haplotypes and the correlation to sampling effort (number of sites). These sample-based curves for North American (▲) and European (■) haplotype diversity from Wares et al (2002, Appendix) demonstrate a clear asymptote in North America and a linear relationship of haplotypes to sampling effort in Europe. These results suggest that with continued sampling, more haplotypes will be found in Europe, while for North America, the number of haplotypes found with continued sampling is levelling off.

Source: adapted from Wares and Blakeslee (2007); analysis performed using EstimateS (Colwell, 2006)

infecting the snail *Batillaria attramentaria* (Asian horn snail), a known introduction to the US Pacific Coast, possessed a handful of haplotypes that were absent in its native population in Asia. If this species was not known to be invasive, it might be presumed that the unshared haplotypes found in North America were endemic to that region since new mutations yielding novel haplotypes would not be expected to arise in the relatively short time that had elapsed since the snail was introduced to the US Pacific coast. The authors concluded that insufficient sampling was probably the explanation, suggesting that the existence of unique haplotypes does not necessarily imply endemism *a priori* (Chapman et al, in press).

Wares et al (2002) used the MDIV program (Nielsen and Wakeley, 2001) to estimate the divergence time between the European and North American populations based on their sequence data. This analysis revealed the maximum likelihood divergence estimate to be 23.2 thousand years ago (kya), with a low-bounded 95 per cent confidence interval of 16.1 kya. However, a complication with

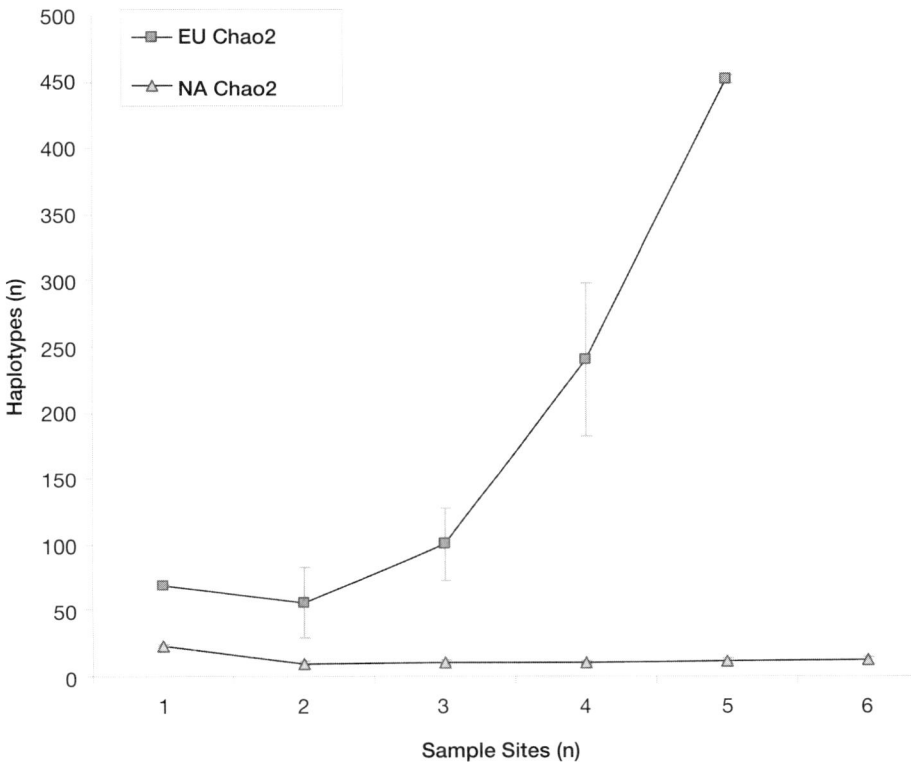

Figure 2.2 *Expected number of haplotypes in Europe versus North America per sampling effort from the Wares et al (2002) haplotype dataset*

Notes: The total number of haplotypes expected in each population was found using the Chao2 estimator, which predicts the eventual asymptote in haplotype richness for a particular population (Gotelli and Colwell, 2001) by including the effects of rare species on the total haplotype richness. This figure portrays the expected number of haplotypes (n) as a function of sampling effort (number of sample sites) for Europe (EU) and North America (NA). At the maximum sampling effort in each population, the Chao2 estimator predicted a total of 450 haplotypes in Europe (■) versus 12 in North America (△), which suggests that with continued sampling, Europe is expected to have 38 times more haplotypes than North America.

Source: adapted from Blakeslee (2007); analysis performed using EstimateS (Colwell, 2006)

Wares et al's (2002) divergence estimate is that the MDIV analysis assumes equal effective population sizes (N_e) for the two populations being compared. Wares et al's (2002) mitochondrial DNA sequence data violate this assumption (Chapman et al, 2007). When Chapman et al (2007) calculated N_e using MDIV for each of the two populations separately (Europe: 60 sequences; North America: 57 sequences), the maximum likelihood N_e was found to be approximately five times larger for Europe than for North America. Because the N_e is part of the calculation for the divergence estimate between the two populations (t = T × $2N_e$, where t = total divergence time in generations, T = a scaled divergence time determined from the sequence data using a maximum likelihood plot, and N_e = effective population size; Nielsen and Wakeley 2001), a great disparity in N_e between populations, where one population is more divergent and has a larger N_e than the other, will result in an inflated divergence estimate between the two populations. Europe clearly has greater diversity than North America, which suggests that the MDIV analysis was inappropriate for Wares et al's (2002) dataset (Chapman et al; Cunningham, in press). Chapman et al (in press) therefore employed an isolation with migration (IM) analysis (Hey and Nielsen, 2004) to calculate the divergence between the two populations for Wares et al's (2002) dataset. IM directly estimates the divergence time (t = t/μ, where t = total years of divergence time, t = the time parameter determined from the sequence data, and μ = the gene substitution rate; Hey and Nielsen, 2004) from the sequence data and therefore does not require the assumption for equal N_e between the two populations. Using IM, Chapman et al (in press) found that divergence estimates were much lower than those derived by Wares et al (2002) from divergence estimates using MDIV (the MDIV estimates were as much as 24 times greater than the IM estimates), probably due to inappropriate assumptions in the MDIV analysis. Furthermore, Chapman et al (in press) argued that the unique North American haplotypes that Wares et al (2002) assumed to be endemic would affect the divergence estimates calculated by both MDIV and IM. Thus, original divergence estimates based on the Wares et al (2002) mitochondrial DNA dataset were probably inflated.

The molecular evidence gathered to date has been debatable or unclear, with two similar investigations (Berger, 1977; Morris, 1979) coming to opposite conclusions, and the mitochondrial DNA dataset compiled by Wares et al (2002) based on too small a sample to offer convincing results (Chapman et al, 2007). A further molecular technique, amplified fragment length polymorphisms, has been used by Wares and Blakeslee (2007) to compare European and North American *Littorina littorea*. This method explores numerous loci, rather than just one or two, and despite providing evidence to suggest that the *L. littorea* was a recent introduction to North America (of 120 loci, none were fixed in either population), it also generated unexpected artefacts that obscured the findings.

Given that the molecular and other forms of evidence have been either inconclusive, ambiguous or both, Blakeslee (2007) developed new lines of enquiry to resolve the cryptogenic status of the *L. littorea*. Initially, she examined the species richness of parasitic trematodes infecting *L. littorea* and two co-occurring congeners, *L. saxatilis* and *L. obtusata*, both considered native throughout the North Atlantic. Only *L. littorea* possessed significantly fewer trematode species in North America, and all three *Littorina* sp. North American trematodes were a nested subset of Europe. Furthermore, using mitochondrial DNA sequencing, Blakeslee discovered that *Cryptocotyle lingua*, the most common trematode parasite of *L.*

littorea, exhibited a significant reduction in haplotype diversity in North America compared to Europe, consistent with signatures of a recent founder effect. She then sequenced nearly 400 *L. littorea* snails using a neutral mitochondrial marker in both populations to determine whether North American *L. littorea* displayed signatures of an introduction. This yielded several indications of a recent founder event for the snail in North America, including a vast reduction in overall genetic diversity in North America compared to Europe, and a divergence estimate within the time-frame of European settlement of North America (Blakeslee, 2007).

In sum, Blakeslee's (2007) results indicate a recent introduction for *L. littorea*, while her parasite analyses confirm other scientific evidence suggesting an earlier natural expansion from Europe to North America for congener snails, *L. saxatilis* and *L. obtusata*.

Conclusions and significance

This corroborative and novel evidence – lower trematode species richness in North American *Littorina littorea*; founder effect signatures in *L. littorea*; and founder effect signatures in an associated trematode parasite – when taken in concert with past historical and ecological evidence, are indicative of a recent introduction of *L. littorea* to North America (Blakeslee, 2007). Moreover, this introduction was probably human-mediated for three main reasons. First, it is unlikely that *L. littorea* crossed the ocean naturally since it is absent from North Atlantic islands such as Iceland and Greenland, which are believed to have served as stepping stones for several intertidal species, such as *L. saxatilis* and *L. obtusata* (Johannesson, 1988; Ingolfsson, 1992). Second, *L. littorea* has been associated with several potential human mechanisms of introduction, including the conveyance of ballast rock and as a human food source (Reid, 1996; Steneck and Carlton, 2001). Third, the rapid, sequential southwards movement of *L. littorea* is a typical occurrence for a recently introduced and rapidly expanding non-native species (Steneck and Carlton, 2001).

The resolution of this 150-year debate is significant for many reasons. It not only enhances understanding of model invasions and the impacts of non-indigenous species on natives; it also provides new tools for the resolution of cryptogenic histories. In addition, this work demonstrates that species invasions are not always singular in nature but can include associated organisms like parasites (such as *Cryptocotyle lingua*). Accordingly, knowledge of cryptogenic histories is important not only at the species level, but may also reveal community-wide impacts. Further research into the resolution of cryptogenic histories is therefore necessary for a holistic understanding of interactions among species (including species with questionable invasion histories) in communities around the world.

Acknowledgements

I thank I. Altman, T. Backeljau, M. Blakeslee, J. Bolster, S. Brawley, D. Brazeau, J. Byers, K. Carleton, J. Carlton, J. Chapman, R. Coleman, L. Collins, L. Curtis, F. Cuthbert, M. Donovan, A. Figueroa, A. Fowler, A. Freeman, G. Goldsmith, J. Grahame, B. Griffen, A. Houghton, R. Houghton, T. Huspeni, M. Janamma, K. Jensen, K. Johannesson, L. Johnson, A. Kintner, T. Kocher, M. Krakau, J. Lee, M. Lesser, T. Maguire, J. Meyer, L. Page, J. Phelan, R. Poulsen, R. Ramsay, A. Rosenberg, D. Starkey, C. Suckling, V. Taibe, S. Teck, F. Viard, J. Wares, S. Williams and D. Zdankoweicz for field and/or laboratory assistance, insights, advice and/or manuscript comments. Support provided by NSF (OCE-0503932), NH Sea Grant, USDA Hatch, and the UNH Vice President for Research Discretionary Fund to J. Byers. A. Blakeslee was supported by the Sloan Foundation History of Marine Animal Populations (HMAP) and small grants from the American Malacological Society, Sigma Xi, Lerner Gray, UNH Center for Marine Biology, UNH Zoology Department and UNH Graduate School. I am also grateful to two anonymous referees for their helpful comments.

References

Bequaert, J. (1943) 'The genus *Littorina* in the western Atlantic', *Johnsonia*, vol 1, pp1–27

Berger, E. (1977) 'Gene-enzyme variation in three sympatric species of *Littorina*. II. The Roscoff population, with a note on the origin of North American *L. littorea*', *Biological Bulletin*, vol 153, pp255–264

Bertness, M. (1984) 'Habitat and community modification by an introduced herbivorous snail', *Ecology*, vol 65, pp370–381

Binney, W. G. (1863) 'North American conchology previous to year 1860. Part I. American authors', *Smithsonian Miscellaneous Collections*, vol 5, pp1–650

Bird, J. (1968) '*Littorina littorea*: Occurrence in a northern Newfoundland beach terrace, predating Norse settlements', *Science*, vol 159, p114

Blakeslee, A. M. H. (2007) 'Resolving the 150 year debate over the ecological history of the common periwinkle snail, *Littorina littorea*, in northeast North America', PhD thesis, University of New Hampshire, Durham, US, pp1–134

Brenchley, G. and Carlton, J. (1983) 'Competitive displacement of native mud snails by introduced periwinkles in the New England intertidal zone', *Biological Bulletin*, vol 165, pp543–558

Byers, J. E. and Pringle J. M. (2006) 'Going against the flow: Retention, range limits and invasions in advective environments', *Marine Ecology Progress Series*, vol 313, pp27–41

Carlton, J. (1982) 'The historical biogeography of *Littorina littorea* on the Atlantic coast of North America, and implications for the interpretation of the structure of New England intertidal communities', *Malacological Review*, vol 15, p146

Carlton, J. (1996) 'Biological invasions and cryptogenic species', *Ecology*, vol 77, pp1653–1655

Chapman, J. W. and Carlton, J. T. (1991) 'A test of criteria for introduced species – the global invasion by the isopod *Synidotea–Laevidorsalis* (Miers, 1881)', *Journal of Crustacean Biology*, vol 11, pp386–400

Chapman, J. W., Carlton, J. T., Bellinger, M. R. and Blakeslee, A. M. H. (2007) 'Premature refutation of a human-mediated marine species introduction: The case history of the marine snail, *Littorina littorea*, in the Northwestern Atlantic', *Biological Invasions*, vol 9, pp737–750

Chapman, J. W., Carlton, J. T., Bellinger, M. R. and Blakeslee, A. M. H. (in press) 'Parsimony dictates a human introduction: On the use of genetic (and other) data to distinguish between the natural and human-mediated invasion of the European snail *Littorina littorea* in North America', *Biological Invasions*

Clarke, A. (1963) 'Supplementary notes on Pre-Columbian *Littorina littorea* in Nova Scotia', *The Nautilus*, vol 77 pp8–11

Clarke, A. H. and Erskine, J. S. (1961) 'Pre-Columbian *Littorina littorea* in Nova Scotia', *Science*, vol 134, pp393–394

Cohen, A. N. and Carlton, J. T. (1998) 'Accelerating invasion rate in a highly invaded estuary', *Science*, vol 279, pp555–558

Colwell, R. K. (2006) *EstimateS: Statistical Estimation of Species Richness and Shared Species from Samples. Version 8.0. User's Guide and Application*, http://purl.oclc.org/estimates, accessed 15 September 2007

Coyer, J. A., Hoarou, G., Skage, M., Stam, W. T. and Olsen, J. L. (2006) 'Origin of *Fucus serratus* (Heterokontophyta: Fucaceae) populations in Iceland and the Faeroes: A microsatellite-based assessment', *European Journal of Phycology*, vol 41, pp235–246

Cunningham, C. W. (in press) 'How to use genetic data to distinguish between natural and human-mediated introduction of *Littorina littorea* to North America', *Biological Invasions*

Ganong, W. (1886) 'Is *Littorina littorea* introduced or indigenous?', *American Naturalist*, vol 20, pp931–940

Gotelli, N. J. and Colwell, R. K. (2001) 'Quantifying biodiversity: Procedures and pitfalls in the measurement and comparison of species richness', *Ecology Letters*, vol 4, pp379–391

Gould, A. A. (1851) 'Mollusca and shells with an atlas of plates', in C. Sherman (ed) *United States Exploring Expedition, during the years 1838, 1839, 1840, 1841, 1842*, Philadelphia, PA

Grosberg, R. and Cunningham, C. (2000) 'Genetic structure in the sea: From populations to communities', in M. Bertness, S. Gaines and M. Hay (eds) *Marine Community Ecology*, Sinauer Associates Inc, Sunderland, MA, pp161–84

Hey, J. and Nielsen, R. (2004) 'Multilocus methods for estimating population sizes, migration rates and divergence time, with applications to the divergence of *Drosophila pseudoobscura* and *D. persimilis*', *Genetics*, vol 167, pp747–760

Ingolfsson, A. (1992) 'The origin of the rocky shore fauna of Iceland and the Canadian Maritimes', *Journal of Biogeography*, vol 19, pp705–712

Johannesson, K. (1988) 'The paradox of Rockall: Why is a brooding gastropod (*Littorina saxatilis*) more widespread than one having a planktonic larval dispersal stage (*L. littorea*)?', *Marine Biology*, vol 99, pp507–513

Johannesson, K. (1992) 'Genetic variability and large scale differentiation in two species of littorinid gastropods with planktotrophic development, *Littorina littorea* (L.) and *Melarhaphe* (*Littorina*) *neritoides* (L.) (Prosobranchia: Littorinacea), with notes on a mass occurrence of *M. neritoides* in Sweden', *Biological Journal of the Linnaean Society*, vol 47, pp285–299

Kraeuter, J. (1976) 'Offshore currents, larval transport, and establishment of southern populations of *Littorina littorea* Linne along the US Atlantic Coast', *Thalassia Jugoslavica*, vol 10, pp159–170

Ledig, F., Conkle, M., Bermejo-Velazquez, B., Eguiliz-Piedra, T., Hodgskiss, P., Johnson, D. and Dvorak, W. (1999) 'Evidence for an extreme bottleneck in a rare Mexican pinyon: Genetic diversity, disequilibrium, and the mating system in *Pinus Maximartinezii*', *Evolution*, vol 53, pp91–99

Lindroth, C. H. (1957) *The Faunal Connections between Europe and North America*, John Wiley & Sons, New York

Lubchenco, J. (1983) '*Littorina* and *Fucus*: Effects of herbivores, substratum heterogeneity, and plant escapes during succession', *Ecology*, vol 64, pp1116–1123

Lubchenco, J. and Menge, B. (1978) 'Community development and persistence in a low rocky intertidal zone', *Ecological Monographs*, vol 48, pp67–94

Miura, O., Torchin, M. E., Kuris, A. M., Hechinger, R. and Chiba, S. (2006) 'Introduced cryptic species of parasites exhibit different invasion pathways', *Proceedings of the National Academy of Sciences of the United States of America*, vol 103, pp19818–19823

Moore, H. (1937) 'The biology of *Littorina littorea*. Part I. Growth of the shell and tissues, spawning, length of life and mortality', *Journal of the Marine Biological Association of the United Kingdom*, vol 21, pp721–742

Morris, S. R. (1979) 'Genetic variation in the Genus *Littorina*', PhD thesis, University of Wales, UK

Nei, M. and Li, W. H. (1979) 'Mathematical model for studying genetic variation in terms of restriction endonucleases', *Proceedings of the National Academy of Sciences of the United States of America*, vol 76, pp5269–5273

Nielsen, R. and Wakeley, J. (2001) 'Distinguishing migration from isolation: A Markov chain Monte Carlo approach', *Genetics*, vol 158, pp885–896

Packard, A. S. (1870) 'Conchological notes', *American Naturalist*, vol 4, p250

Reid, D. G. (1996) '*Littorina littorea*', in D. G. Reid (ed) *Systematics and Evolution of Littorina*, The Ray Society, Andover, UK, pp95–12

Ruiz, G., Rofonoff P., Carlton J., Wonham M. and Hines, A. (2000) 'Invasion of coastal marine communities in North America: Apparent patterns, process, and biases', *Annual Review of Ecology and Systematics*, vol 31, pp481–531

Spjeldnaes, N. and Henningsmoen, K. E. (1963a) '*Littorina littorea*: An indicator of Norse settlement in North America?', *Science*, vol 141, pp275–276

Spjeldnaes, N. and Henningsmoen, K. E. (1963b) 'Reply to A. H. Clarke', *Science*, vol 142, p1022

Steneck, R. and Carlton, J. (2001) 'Human alterations of marine communities: Students beware!', in M. Bertness, S. Gaines and M. Hay (eds) *Marine Community Ecology*, Sinauer Associates Inc, Sunderland, MA, pp445–46

Verrill, A. E. (1874) 'Results of dredging', *American Journal of Science*, vol 7, pp131–138

Wagner, F. (1977) 'Palaeoecology of marine pleistocene mollusca, Nova Scotia', *Canadian Journal of Earth Sciences*, vol 14, pp1305–1321

Wares, J. P. and Blakeslee, A. M. H. (2007) 'AFLP data provide poor resolution to the *Littorina littorea* puzzle', *Marine Biology Research,* vol 3, pp168–174

Wares, J. and Cunningham, C. (2001) 'Phylogeography and historical ecology of the North Atlantic intertidal', *Evolution*, vol 55, pp2455–2469

Wares, J., Goldwater, D., Kong, B. and Cunningham, C. (2002) 'Refuting a controversial case of a human-mediated marine species introduction', *Ecology Letters*, vol 5, pp577–584

Willis, J. (1863) 'On the occurrence of *Littorina littorea* on the coast of Nova Scotia', *Transactions of the Nova Scotia Institute of Natural Science*, vol 1, pp88–90

Yamada, S. and Mansour, R. (1987) 'Growth inhibition of native *Littorina saxatilis* (Olivi) by introduced *L. littorea* (L.)', *Journal of Experimental Marine Biology and Ecology*, vol 105, pp187–196

3

Social Conflict, Over-Fishing and Disease in the Florida Sponge Fishery, 1849–1939

Loren McClenachan

During the late 19th century, poor coastal fishermen in Florida prospered from the sale of millions of pounds of bath sponges to the North American metropolises of New York, Chicago and St Louis. 'Sponger money' was said to 'ever flow' (Sweeting, undated) and fishing for marine sponges formed the basis of entire local economies along the coast. Money flowed into the pockets of small-scale fishermen who transformed living sponges into Florida's most valuable marine commodity. Until 1905, the banks of marine sponges in the reefs and hard-bottom environments of the Florida Keys were partially protected from over-fishing by the awkwardness of the fishing gear. Key West spongers used long-handled rakes to grope for sponges in the shallows, leaving deep-water sponges to grow and reproduce.

However, this rosy picture of prosperity faded for these 'traditional' boat-based fishermen during the early years of the 20th century. The declining health of marine sponge populations – together with an unexpected combination of technological innovation, war and discovery – initiated a dramatic and permanent change in the industry. Mediterranean sponge fishermen revolutionized the Florida fishery by introducing diving technology and shifting the centre of the fishery north to the Gulf Coast. Commercial divers from Greece could take sponges from water depths unimaginable to the Key West spongers, and their ability to access pristine sponge banks and more efficient sponge curing techniques helped them to outcompete traditional fishermen. In 1905, Greek divers began to arrive in Florida in gold rush proportions, and by 1910 they outnumbered the Key West fishermen (Moore, 1910). Social conflict, intensified by an increasingly over-fished resource, characterized the Florida sponge fishery during the early 20th century.

This chapter explores the conflict between Greek fishermen in Tarpon Springs and the Key West fishermen who were largely of Bahamian descent. It argues that the last three decades of the fishery were characterized by market-driven resource competition between two ethnically different groups of fishermen. This social conflict was heightened by overstressed sponge resources that eventually disappeared entirely due to a synergistic relationship between over-fishing and disease in the marine environment. Two themes central to historical marine ecology emerge from the story of sponge fishing in Florida. First, changes in technology can revolutionize a fishery and temporarily breathe life into a dying industry, but ultimately push a resource to commercial and ecological extinction. Second, the interaction between over-fishing and disease in the marine environment is not an exclusively modern phenomenon. Understanding synergistic stresses on structural elements of the marine community is central to environmental conservation, particularly in coral reef ecosystems that have been plagued by various unknown diseases.

At its peak, the fishery in the northern Caribbean removed 47 million pounds (lb) of live sponge annually, and over-fishing was evident in both quantitative and qualitative descriptions of the fishery in Florida. A resource manager during the early 20th century observed that there were so many boats and so few sponges left on the reefs that the spongers could 'hardly wield their poles without collision' (Moore, 1910, p440). Despite efforts on the part of the Key West fishermen to limit the activity of divers, high levels of fishing continued. Disease drastically reduced the remaining commercial sponge populations in 1939, leaving the fishermen in a 'pitifully destitute condition' (Barbour, 1945, p113). The era of sponge fishing lasted less than a century, but was critical to early economic development and has left its mark both on the culture of south Florida and the ecology of the reefs. Those reefs have yet to recover from the annihilation of commercial sponge species.

Marine sponge biology and uses

Alive on a coral reef, marine sponges superficially resemble the coral itself: large living masses shaped like barrels, long asymmetrical fingers or compact spheres. Fish and lobsters congregate around them. Hawksbill turtles sit on the seafloor and tear through their thick skin and hard skeleton to feed on the living matter inside. Biologically, they are nothing like corals or any other animal. The most basic of multicellular organisms, sponges are simple in their structural design – essentially, a series of tiny canals that direct water from the surrounding environment to deep within the sponge body where nutrients are extracted; but they host a complicated assortment of biological symbioses and chemical compounds. Bacteria and other micro-organisms that can resist digestion often inhabit the sponge so that more than half of the living tissue in a sponge can be comprised of a cocktail of chemically defended micro-organisms (Lee et al, 2001).

As living water filters, marine sponges do their job better than any other animal in the ocean. Oysters and other bivalves use small hairs to grab relatively large pieces of suspended organic matter from the water; but sponges assimilate particles ranging over an order of magnitude in size and can subsist entirely on the smallest bacteria (Stuart and Klumpp, 1984). The nutritional value of these tiny bacteria is low so that sponges need to be remarkably efficient both at extracting cells from the water and pumping seawater through their canals. They have perfected the art of water filtration; tropical sponges can remove over 90 per cent of the bacteria present in the water and can cycle through up to 20,000 times their own volume in a 24-hour period (Reiswig, 1971, 1974). In the Caribbean, where sponges abound, they effectively clean the bacteria in the water surrounding the reef several times per day (Reiswig, 1974).

Well over 1000 genera of marine sponges exist; but only two have commercial value. Species of *Hippospongia* and *Spongia*, which are found on reefs, hard-bottom communities and sandy bottoms at depths between 2m and 60m, have been fished for centuries because their naturally soft and absorbent dried skeletons had endless uses before the invention of modern plastics. Early Europeans made sponges into padding for their helmets, lightweight and transportable drinking devices and even early municipal drinking water filters (Anonymous, 1939; Hall and Dietrich, 2000). Unlike most sponges, the skeletons of commercial sponges are not made of chalk-like calcium carbonate or glass-like silica spicules. Instead, once the living tissue is removed, all that remains of *Hippospongia* and *Spongia* is an ornate matrix of spongin: a soft, elastic protein-based fibre.

In the century before synthetic sponges, personal hygiene, house and car cleaning, medical surgery and the fine arts all required natural sponges. Art historians can date pieces of pottery by the type of sponge used to apply the glaze, while natural sponges played a role in women's reproductive freedom. Doused with 'antiseptic', they were commonly marketed as underground contraceptives during the late 19th century when planning parenthood was highly controversial (Brodie, 1994). Commercial grades of sponge correspond to different species (see Table 3.1). The most valuable sponges – the wool and velvet sponges – were prized for their soft texture. Species of *Spongia* supplied lower-grade sponges, including the wire and hardhead sponges, whose names imply that they were more frequently used for house cleaning than the more delicate tasks of applying pottery glaze or preventing pregnancy.

Manning the Key West sponge fishery from Bahamian waters, 1839–1890

Although politically part of the United States' newly acquired state of Florida, 19th-century Key West was in reality a tiny Caribbean island outpost with fewer than 1000 residents, accessible only by boat and isolated from the rest of the nation. Its modest economy was local and marine based, with few products for export, the despatch of live fish to Havana occupying more than one third of Key West's boat

Table 3.1 *Species targeted in the Florida sponge fishery*

Trade Name	Value	Species
Wool	Very high	*Hippospongia lachne*
Velvet	Very high	*H. gossypina*
Yellow	High	*Spongia barbara*
Grass/Vase	Varies	*S. graminea*
Wire	Low	*S. sterea*
Glove	Very low	*S. cheiris*
Reef	Very low	*S. obliqua*
Hardhead	Very low	*S. agaracina*

Source: data derived from Moore (1910)

traffic during the 1830s (KWCHR, 1837–1840). The discovery of sponges that could be dried and shipped to northern cities was therefore a boon to the local economy and a resource that poor men could easily turn into money. In the first years of the fishery, spongers simply waded into 'shoal waters in the immediate vicinity of Key West' where sponges were pulled by hand (Moore, 1910, p435). Fishermen tore sponges from the ocean floor, brought them to the surface, squeezed them hard so that the 'gurry' – the living tissue, associated bacteria and organic material – oozed out of the animal. The dead skeleton was then left to soak, before being dried and cut into geometric shapes for sale.

As the industry developed, small boats became necessary to access sponges in deeper water (see Figure 3.1), and men who were familiar with such sponge fishing vessels began to arrive in the Florida Keys from the nearby Bahamas. The Bahamian sponge fishery was the first in the Caribbean and had existed for a decade before the first Florida sponge was exported in 1849. Experienced Bahamian spongers were drawn to the economic opportunities in Florida and driven from the Bahamas by poor conditions that existed there for black labourers. In the Bahamas, sponge fishermen, like workers on the colony's pineapple plantations or sisal farms, were typically ex-slaves who owned limited personal property and had little chance of improving their income and their families' welfare. Despite the seemingly endless boatloads of marine sponges pulled from the reef and exported during the mid 19th century, workers remained poor. White men in Nassau controlled the business and spongers were remunerated by shares, which were sometimes paid in kind – usually supplies – rather than in cash (Johnson, 1988). The decline of the practice of salvaging shipwrecks – an industry that had employed nearly half of the Bahamas' able-bodied men during the 1850s (Albury, 1975) – and the collapse of the pineapple and sisal agriculture in the 1870s left large numbers of Bahamian workers reliant on a sponge industry that provided little financial reward (Johnson, 1988).

The sponge industry in Florida offered employment that experienced Bahamian boatmen badly needed and for which they were well qualified (see Figure 3.2). The

Figure 3.1 *Schooners at Key West Sponge Docks, 1902*

Source: image from the Monroe County Public Library photograph collection

first Bahamian settlers in Key West during the 1840s were financially independent fishermen and wreckers; but as industry developed in Key West, opportunities for labourers increased. By 1885, six schooners were engaged exclusively in transporting Bahamians from Nassau to Key West (Blake, 1885). As word spread that there were 'places in the world where not only a high rate of wage is paid, but the people get paid in cash' (Powles, 1888, p88), Bahamians came to Key West in droves, contributing to the doubling of the island's population between 1880 and 1890. These migrants frequently brought their entire families or, in one case, 'a whole collection of families', which expressed their 'disgust with the Colony' and left 'their substantial houses' in Rock Sound for south Florida. Like most of their compatriots, these 40 or so Bahamians did not return (Haynes-Smith, 1896). Accordingly, by 1890, Bahamians comprised over one quarter of the 20,000 residents of Key West – and a majority of its 2000 spongers (Brice, 1897).

Bahamian boatmen were attracted to Key West by the improved economic opportunities afforded by the sponge industry, and black fishermen further found opportunity for advancement in the burgeoning city. The racial dynamics of

Figure 3.2 *Sponge workers trimming sponges, Nassau, the Bahamas, circa 1932*

Source: C. M. Yonge (photo credit), Natural History Museum, London, UK

Florida's sponge industry were complex, but black men – particularly those with families – encountered a degree of acceptance within the south Florida fishing community (Goode and Collins, 1887). Race relations in south Florida during the mid 19th century were influenced as much by Cuban society as that of the American south or the Bahamas. Cubans considered south Florida to be 'so much a part of their own country' and fishermen and labourers came and went through the porous national border (James, 1998, pp237). Cuba, the Bahamas and south Florida all had slightly different racial etiquettes, and many black men found a respectable place in this international maritime frontier. During the 1890s, black fishermen were central to the sponge industry and a high degree of racial mixing occurred aboard sponge boats. Some 65 per cent of Key West spongers were black and most boats contained crews of both black and white spongers (Cobb, 1902). As time passed, black spongers were among those who successfully sought sanctions from the legislature to curb the activities of immigrant Greek divers.

PROSPERITY AND UNINTENTIONAL CONSERVATION, 1890–1905

Evidence of sponging was everywhere in late 19th-century Key West, where backyards and boat-yards were piled high with drying sponges (see Figure 3.3). By the turn of the century, total annual export was in the range of several hundred thousand pounds of dried sponges, the equivalent to over 1 million pounds of living animals. These sponges supplied more than half of the total American market. The optimism for the future of the industry was reflected in popular culture; the song 'Sponger Money Never Done' was sung both in the Bahamas and Key West during the 19th century (Rolle, 1940; Dean, 1981; Sweeting, undated):

> ... *Look in my trunk and see what's there, sponger money,*
> *One hundred dollars was my share, sponger money,*
> *I'm gonna take away your woes, sponger money,*
> *I'm gonna buy you fine new clothes, sponger money.*
> *Then when we go out on the street, sponger money,*
> *You'll be lookin' nice and neat, sponger money,*
> *Then all the boys will envy me, sponger money,*
> *Then all the girls will fall for me, sponger money.*
> *Money don't make me you know, sponger money,*
> *Sponger money ever flow, sponger money,*
> *Tell ev'rybody in town, sponger money,*
> *Me and my gal gon' dance 'em down, sponger money.*
> *Sponger money never done, sponger money.*

Figure 3.3 *Sponges drying on a platform at the rear of sponge house, Key West, December 1921*

Source: R. E. Coker (photo credit), NARA, College Park

The days of wading into shallow water and collecting sponges from the reefs were gone, however, and it became necessary to travel to 'deeper and deeper water year after year' to find marketable sponges (Brice, 1897, p267). When sponges in the shallows disappeared, spongers adopted the makeshift technology of long-handled rakes with hooks on the ends and oil from nurse shark livers, which they poured on the water's surface on choppy days to smooth the surface ripples. Spongers looked through the clear water of the Keys, located banks of living sponges on the sea floor, and pulled them with their rakes. In 1888, marking the further depletion of shallow-water sponges, a new kind of hook was developed to better access the sponges in water deeper than 30 feet (9m) (Brice, 1897). By the 1890s, shallow-water sponges were nearly 'fished out' (Brice, 1897, p267) and spongers had been forced out to sea, 'until the maximum depth in which sponging is possible' had been reached (Smith, 1898, p232).

The Key West fleet was comprised of 100 to 200 small sailing vessels worked by crews of five men or less. These vessels typically embarked on week- to month-long fishing trips, during which two-man dinghies were despatched over reefs (see Figure 3.1). The limited technological advances in the industry served primarily to extend the area accessible to sponge boats into deeper water; but without diving gear, expansion was limited. Before 1905, the biggest technological innovation was the 'water telescope', essentially a glass-bottom bucket that replaced shark liver oil and improved the sponger's ability to see the bottom. Because access to sponges was ultimately limited by depth, sponging became a profession in which individual skill made a great difference. In deep water – depths of over 40 feet (12m) – just one third of the spongers had 'sufficient strength, keenness of sight and skill with the pole to work successfully' (Moore, 1910, p439). Prosperity was thus linked to specific talents, and as the fishery became more developed, competition fiercer and the sponges more depleted, many spongers found the industry to be less welcoming than it had been when they arrived from the Bahamas.

Just as the shallow banks within wading distance of Key West had been depleted, the sponges at depths of 30 to 40 feet (9m to 12m) were proving finite. Declines in size and quality of sponges were noticed by the 1890s, and the large sponges that had frequently occurred in the past were 'rarely, if ever, seen' by the end of the century (Brice, 1897, p268). Spongers themselves noted declines in yields. A typical trip in the 1880s brought in 1000 to 1500 bunches of sponges; by the late 1890s, 500 bunches were considered a good cargo (Brice, 1897). Sponge grounds were simply 'much less productive than formerly', a predicament acknowledged by 'practically everyone who is in a position to express an intelligent opinion' (Smith, 1898, p232). Inshore grounds were abandoned and smaller catches per vessel were noted, while yields primarily comprised the smallest sponges (Smith, 1898).

Intensive fishing in shallow water drove such decline; but protection of stocks was granted by the clumsy nature of the fishing gear and the inaccessibility of sponges in deeper water. The most valuable sheepswool sponges were commercially extinct in near-shore areas by the turn of the century, but they were markedly more abundant in waters deeper than 40 feet (12m), illustrated by the fact that experienced captains were 'in the habit of running offshore from time to time to take a sight' of the bottom with their water glasses. If conditions were favourable and the water was clear, they could easily 'take more sponges in a day than they could in a week on the overworked inshore grounds' (Moore, 1910, p439). Sponges had been spotted far deeper: one was brought up from 100 feet (30m) on a fishing line and several men reported banks as deep as 120 feet (37m) – but these were impossibly deep (Moore, 1910). As long as abundant deep-water sponges remained untouched, they acted as a *de facto* reserve population that helped to reseed the shallow-water sponge banks.

TECHNOLOGICAL REVOLUTION AND COMPETITION IN THE MARKETPLACE, 1905–1910

During the early 1870s, a second sponge ground had been discovered in the Gulf of Mexico, on Florida's west coast, and a small number of boats worked on these banks for nearly three decades (see Figure 3.4). The distance back to Key West made large-scale sponging in the gulf impractical, and catches of the few boats that worked there were sold to three dealers in Tarpon Springs. In 1895, just 5 per cent of Florida's sponges were sold in Tarpon Springs (Schroeder, 1924; see Figure 3.5). However, fear of Spanish vessels during the Spanish American War drove Key West fishermen to land their catch in Tarpon Springs, and soon this small settlement became a real rival to Key West (Cobb, 1902). By 1902, sponge fishermen brought nearly 20 per cent of the total catch to sell in Tarpon Springs (Schroeder, 1924). Growth of the sponge market attracted settlers, and among the shore-side workers in Tarpon Springs were Greek men from the Mediterranean sponge-diving industry (see Figure 3.6). Like many divers, they considered fishing sponges with rakes – as was done throughout the Caribbean – to be primitive since this ripped the sponges and resulted in an imperfect product. An investigation of the deep-water sponges in the eastern Gulf of Mexico found enough to warrant establishing a diving company, and the first successful diving season was completed in 1905 (Moore, 1910).

Diving increased the depth at which sponges could be collected from 40 to 180 feet (12m to 55m), making accessible thousands of new sponge banks. The addition of deep-water sponging grounds on the Gulf Coast increased the area open to spongers from 4000 to 10,000 square miles (10,300 to 25,900 square kilometres), stretching from the Keys in the south to Apalachicola, high up on the peninsula (see Figure 3.4). The new diving boats were extremely successful at finding and exploiting rich banks that yielded significant quantities of large, valuable sponges, the like of which had not been landed for half a century. One diving boat was memorialized for collecting 500 bunches of high-quality sponges in five days 'without moving much over a mile' (Moore, 1910, p443). In the first

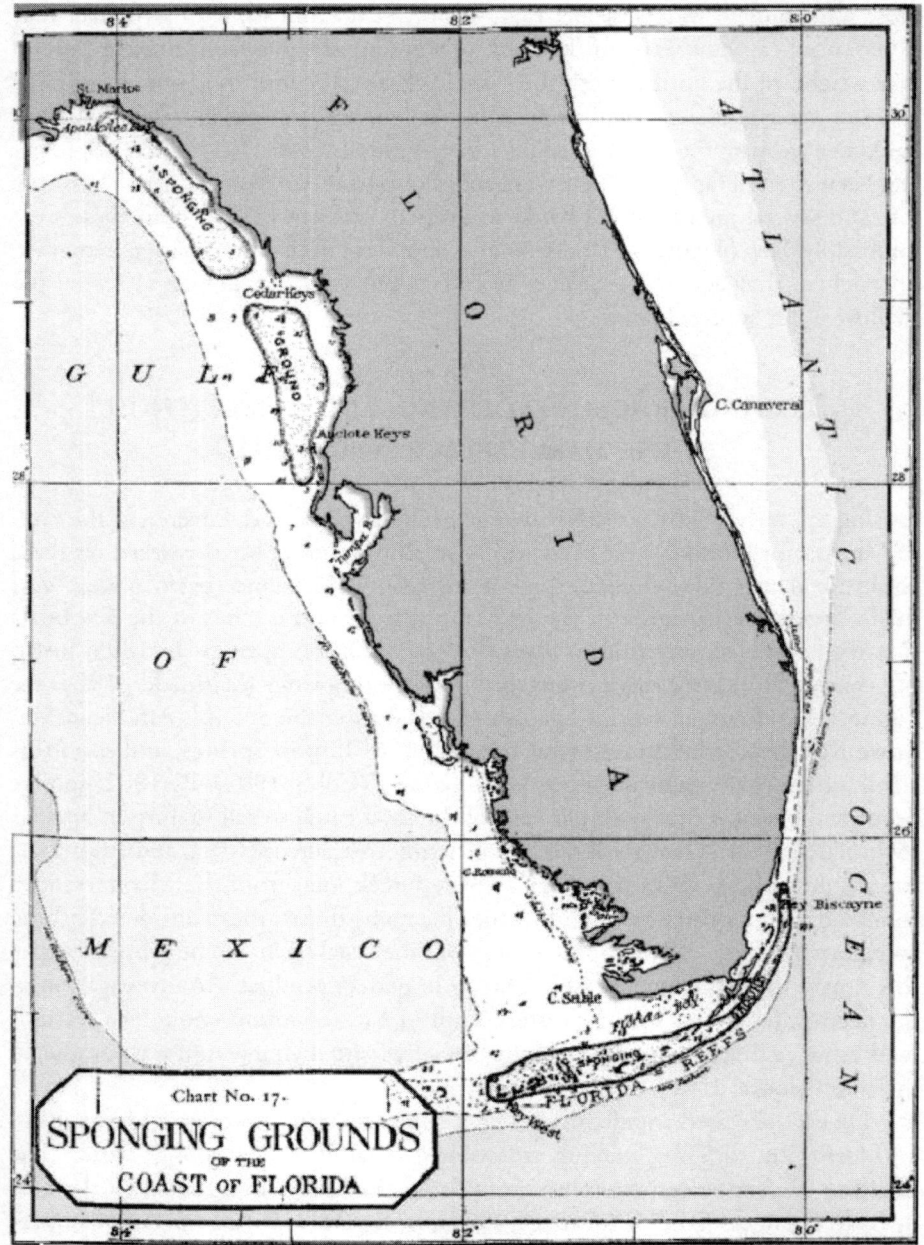

Figure 3.4 *Florida sponging grounds*

Source: adapted from Goode (1887); view colour version at www.hull.ac.uk/oceanspast

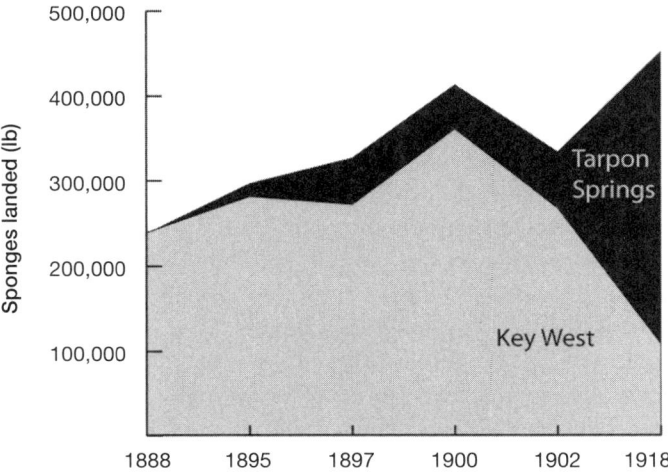

Figure 3.5 *Sponge catches, Key West and Tarpon Springs, 1888–1918 (lb)*

Source: Schroeder (1924)

six weeks of diving in 1906, over US$60,000 worth of sponges were landed, more than 75 per cent of the total annual gulf catch by traditional methods (Cobb, 1902; Moore, 1910). By mid May 1906, there were 50 diving boats employed and 55 more rigged and waiting for crews, 'which could not be supplied despite a great influx of Greek divers and boatmen' (Moore, 1910, pp442–443). The desperate need for divers and the tremendous profit to be made was advertised in New York City and Greece, and hundreds of divers came to harvest the rich crop of sponges in the gulf. Within a year, the population of Tarpon Springs had grown from 700 to 800, with more arriving 'on every train' (Moore, 1910, p443).

The first months of sponge diving in Florida were both successful and profitable, but prices crashed as the supply of sponges from the gulf expanded rapidly. Sponges that were worth $8 to $10 in the opening weeks of 1906 were selling for $3 by May (Moore, 1910). Without market controls, intensive fishing lowered prices; without controls on the fishery, lowered prices increased competition and stimulated more intensive fishing. Divers continued to make profits; but their efficient methods and access to deep-water sponges severely disadvantaged the boat-based spongers. Greek diving boats operated in pristine deep-water sponge beds, where large numbers of high-quality sponges were found. Additionally, they could work on days when the water was choppy, generally stayed out on the grounds longer and cured the sponge catch on board rather than sailing at the end of each week to crawls: shallow-water pens used by spongers and turtle fishermen. Key West boats, in contrast, could only make trips when the water was clear and were confined to the same tired sponge banks that had been worked for decades (Moore, 1910). These practices translated into a measurable economic advantage for the Greek spongers. In 1909, *Filmore*, a Greek schooner with a crew of ten spongers, made US$2180 on a 60-day fishing voyage, whereas a comparable Key West schooner employing 13 spongers made U$1180 on a 42-day trip. These figures suggest that Greek sponging boats earned nearly 70 per cent more per day on the water than the Key West fishing boats working with traditional methods (Moore, 1910).

The Key West spongers initially complained only about the drop in wholesale sponge prices and considered the 'trouble with the Greeks' to be a temporary annoyance rather than a career-ending event. Massive quantities of cheap sponges continued to be available in Tarpon Springs, however, and sponge buyers overlooked Key West markets. This 'cut-throat competition', together with a declining yield from the overworked sponge banks, quickly forced many boat-based spongers out of the industry. Between 1900 and 1908, the number of men in the raking fishery declined by more than 50 per cent (Moore, 1910).

COMPETITION AND VIOLENCE ON THE WATER, 1910–1939

Economic competition characterized the early stages of the diving industry; but as more boats crowded onto the waterfront, competition over access to grounds was not far behind. At first, recognizing that they were newcomers intruding on the Florida sponging grounds, the Greek divers treated the traditional spongers respectfully and confined work to sites far offshore (see Figure 3.6). The deep-water sponges were more valuable; but additional dangers incurred with deep diving made inshore diving desirable. As it became more apparent that the 'the loud threats' of some of the Key West spongers were largely idle, divers became more confident and moved shoreward into the shallow sponge beds (Moore, 1910, p444).

Fierce competition led to changes in the method of fishing as the boat-based spongers attempted to increase their efficiency and stay in business. The Key West men were desperate to find sponges and gathered around productive patches, so that it was common to see them packed so tightly together that the small boats collided (Moore, 1910). In order to make sponging in over-fished areas worthwhile, the spongers changed their method from blindly wielding their poles over known sponge beds to systematically surveying larger grounds and dropping anchors on discovered sponge banks. This method increased productivity in regions where the sponges had become 'too few and scattered for the usual methods to yield results' (Moore, 1910, p440). Finally, in the most depleted areas, vessels began to anchor close to the sponging grounds at night so that they could be first on the grounds in the morning, when the winds were light and the conditions best for work (Moore, 1910).

Greek divers could work as deep as 180 feet (55m); but they preferred to dive on shallow, more heavily fished sponge beds, an irony that revealed the relative inefficiency of the traditional spongers. Despite decades of fishing and declines in

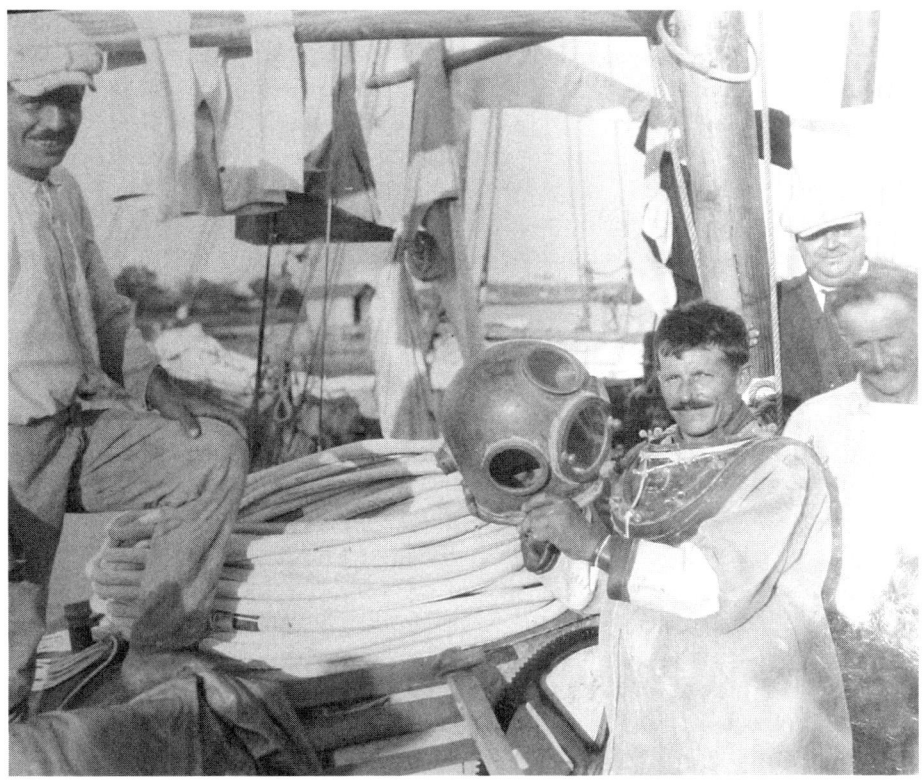

Figure 3.6 *Greek sponge divers, Tarpon Springs, circa 1931*

Source: W. L. Schmitt (photo credit), Smithsonian Institution Archives

yield, sponges still remained in shallow water, hidden in crevices or under ledges. In fact, the largest complaint of the Key West spongers was that Greek divers thoroughly denuded the bottom of all sizes of sponges, leaving fewer to reseed the beds (*KWC*, 8 June 1938). In particular, where divers worked on already depleted grounds, it was clear that the diving industry exhausted sponge populations. Opposition from the boat-based spongers crystallized, and bills were introduced into the state legislature to restrict the use of diving equipment. Among the most convincing arguments for restrictions was that the beds were 'already worked to or beyond their limits of productiveness' and that diving increased 'the draft upon them [and] must be injurious' (Anonymous, 1939, p111). Eventually, three types of bills passed: the diving fishery was closed from May to September, shallow-water and near-shore diving was prohibited, and the minimum size of sponges was increased from 4 to 5 inches (10cm to 13cm) in diameter (Moore, 1910; *KWC*, 30 June 1938).

By the 1920s, the sponge grounds were neatly parcelled and the season divided; but frequent incursions were made. In 1931, 40 miles' (64km) worth of buoys were sought to 'mark the line between diver and hooker territory on the Taylor County battle front' (*FOR*, October 1931, p13). In the same year, 76 divers were fined for violating the shallow-water ban. In the Keys, law enforcement was most vigilant as local authorities tried to catch divers from the gulf violating the law. Raids on non-local boats were frequent, but convictions were rare. In one case in 1930, a boat was raided in the Marquesas, uninhabited islands just west of Key West, and seven men were charged with taking undersized sponges. The conviction was particularly satisfying for local spongers as several of the guilty men were former Key West spongers who had defected to Tarpon Springs to work for a diving operation (*KWC*, 26 June 1930).

Law enforcement officers attempted to keep divers out of shallow water; but angry spongers often took matters into their own hands. Violent confrontations on the waterfront escalated in the Florida Keys and along the Gulf Coast, with angry spongers even resorting to arson. In 1918, Key West spongers set fire to a Greek sponging vessel, the first of four craft to 'be mysteriously destroyed' in the Keys. A schooner 'burned to the water's edge' in 1923 and Greek spongers were warned in the local newspaper to leave the Keys completely (*KWC*, 27 April 1933). A national fisheries trade journal considered such acts of violence to be 'one chapter in a decade or more of warfare' between the sponge divers and boat-based spongers (*FOR*, October 1931, p13).

CONSERVATION ETHICS IN A TIME OF SCARCITY

By 1927, there were just 60 boats and 500 men in the Key West sponge fishery – compared with over 2000 men during the 1880s; but the number of spongers was set to drop even further. In 1938, 40 men in Monroe County called themselves spongers, but there were only five who worked regularly: 'Ben Felton, Nelson Spencer, William Spencer, James Thompson and John Spencer' (*KWC*, 8 June 1938). The Key grounds had been considered exhausted at the turn of the century, and under continued pressure, its sponge banks were almost commercially extinct by the 1930s. Along with over-fishing, blockages in freshwater outflow from the Everglades due to the Key West railroad extension contributed to sponge mortality throughout the Keys (Moore, 1910). As the numbers of spongers and the available sponge supply shrank, Key West fishermen fought to maintain a presence in the fishery by means that were at least moderately conservationist. Many of their economic interests became aligned with a conservation ethic as they pressed for legislation to protect the deep-water sponge banks and their own inefficient techniques.

Limiting technology certainly helped to maintain sponge populations, particularly in deep water, and Key West spongers were fiercely protective of their local rights and suspicious of any type of technological progress that threatened to constrain their rights. They were said to hold antagonistic attitudes towards any kind of innovation, whether it was diving equipment or sponge culture, even though the latter may have aided the recuperation of the beds. Correctly perceiving the privatization of the ocean bottom as a threat to their immediate livelihoods, Key West spongers rejected attempts to experiment with aquaculture. A lack of proactive government mediation caused sponge fishermen to be pitted against businessmen and scientists interested in aquaculture. In one case involving experimental seeding of sponges, more than 100 men protested a court order to not disturb sponges in the shallow bay of Sugarloaf Sound (*KWC*, 13 October 1930). Spongers in the Florida Keys thus clung to a way of life that had been prosperous a generation earlier, but was on the brink of collapse by the 1930s.

Key West spongers who did not leave the business during the 20th century fought to stay relevant, while the Tarpon Springs fishery appeared optimistic and ascendant. Three times more sponges were landed in Tarpon Springs than in Key West in 1918 (Schroeder, 1924; see Figure 3.5) and the sponges were worth up to 90 per cent more per pound than those from shallower water (*KWC*, 14 April 1930). Confident in their huge profits, gulf divers attempted to expand the area fished and the number of divers in the industry. In 1934, they sought a special permit from the US Immigration Department to enable 40 Greek divers to enter the country (*FOR*, April 1934, p12). Divers complained that large numbers of deep-water sponges existed in the Keys that were lost to the industry and tried on several occasions to establish a base in the Florida Keys (*FOR*, May 1938, p10). Intending to circumvent the hostility of Key West spongers, they petitioned for permission to establish themselves in Miami (*KWC*, 10 April 1930).

Optimism in the diving industry fuelled high levels of investment, which ultimately caused the Tarpon Springs fishery to grow beyond the capacity of the sponge resource. The continued influx of divers was unsustainable, particularly as a pathological disregard for conservation laws eroded the strength of the diving industry. The lack of a conservation ethic among divers – particularly with respect to taking undersized sponges – led boat-based spongers to believe that divers would stop at nothing until all the sponges were gathered from Florida. Boat-based spongers in the Keys and Taylor County resented depletion of their sponge beds by divers and confrontations over taking small sponges fuelled turf battles and cries for further diving restrictions (*FOR*, May 1938, p10). In 1935, the federal government sent an officer to enforce size limits in the gulf. Three years later, a state conservation vessel was despatched to Monroe County to enforce the ban on shallow-water diving (*KWC*, 22 September 1938). Despite the presence of a conservation official, every diving boat was said to have 5in sponges among its catch, leading a former president of the Tarpon Springs sponge exchange to urge Washington to take a more aggressive stance on the protection of abused sponge beds in the gulf (*KWC*, 17 November 1938).

Despite a common desire to turn marine sponges into a cash income, the Greek divers and the boat-based spongers differed in their perspective on conservation and on limiting exploitation. To some degree, Key West fishermen were more ecologically benign, simply because of their inability to reach sponges in deeper water. Had they been able to do so, they certainly would have taken more sponges. Unlike Greek divers, however, the Key West fishermen were not reputed to take small sponges habitually, and they appeared to have a better sense of the limits of the sponge resource, at least in the later years of the fishery. Many Greek divers came as single men to Tarpon Springs, either directly from the over-fished Mediterranean sponge banks or from New York City, and therefore had a relatively high degree of mobility. Diving required substantial capital investment and division of labour so that the businessmen running the industry were farther removed from the environment itself (Brice, 1897). In contrast, the Key West sponge industry comprised a collection of small-scale entrepreneurs who knew no waters other than the Caribbean, had very little capital invested and few alternatives if the industry failed. Thus, these fishermen had a stronger connection to the future ecological state of the marine environment. Key West spongers sought to limit the excesses of the divers both to protect themselves and to protect the sponge resource, upon whose profits the city of Key West was built.

Disease, over-fishing and the unravelling of Florida's coral reef ecosystems

In December 1938, spongers on the other side of the Gulf Stream in the Bahamas began to have an odd and disconcerting experience, one which old-timers who had been in the business since boyhood had never witnessed. Instead of pulling up intact sponges, sponge hooks came to the surface with only slivers and strings; the rest of the sponge skeleton had disintegrated (Anonymous, 1939). When sponges were cut open 'rotten and evil smelling tissue' was exposed (Walton-Smith, 1941, p419). The 'mysterious blight' that struck sponge beds in the Bahamas 'quickly reached epidemic proportions', leaving the ocean floor covered with thousands of bleached and rotted sponges (*FTU*, 6 April 1939). By February, 'there remained no sponge-bearing bank unaffected' in the Bahamas (Galtsoff et al, 1939, p807), and Key West fishermen watched their sponge banks with dread anticipation. In March, signs of disease appeared and by May the Key West sponges showed considerable damage from blight. By the end of 1939, the yellow and vase sponges had suffered nearly 100 per cent mortality, while 70 per cent of the valuable sheepswool sponges had been eliminated from the Keys (Galtsoff, 1939). Although overflowing fresh water and local pollution were both blamed, the culprit appeared to be a fungus-like filament, which started in a small area of the sponge and expanded until the entire sponge was consumed (Galtsoff et al, 1939).

Initial fears that the blight would also affect the sponge beds in the gulf went unfulfilled during the year. In May 1939, Tarpon Springs sponge sales were at their highest in five years (*FOR*, May 1939, p9). By early 1940, however, the disease had made its way up the coast and Greek divers observed extensive sponge mortality as deep as 70 feet (12m) (*FTU*, 6 April 1939). By the end of 1940, the remaining sponge fishermen in Florida were out of work. The disease devastated the industry, leaving fishermen 'destitute' (Barbour, 1945, p113). Flooding the market with sponges in 1906 had caused local prices to drop by two-thirds, but wiping out the northern Caribbean sponge banks caused global prices to increase from US$2 to US$30 per lb between 1939 and 1945 (Tierney, 1949).

While the immediate cause of sponge mortality was disease, the fishermen were far from blameless in the disappearance of the stock. Evidence of over-fishing throughout Florida is prevalent, both in descriptive accounts and in fisheries statistics. In 1930, gulf divers complained that there were so few sponges left in the gulf that could be legally taken by diving (*KWC*, 10 April 1930) that they were obliged to travel long distances to find their prey, as far as five times the distance their predecessors had travelled in the first decade of the 20th century. In the early years of the fishery, large vessels in bay grounds had fished 30km to 50km from the shore, and many divers worked in the shallow waters, taking the scraps left behind by the boat-based spongers (Moore, 1910). By 1938, divers had moved far offshore, some going up to 240km from the coast to find sponges. During the 1930s, the best sponges were found only at depths of up to 120 feet (37m) (*FOR*, May 1938, p3). The size of individuals had also declined consistently, 'owing to the intense fishery' that allowed sponges 'no opportunity to grow' (Moore, 1910, p429).

The disease that exterminated Florida's sponge community during the 1930s was intimately linked to over-fishing, both in the long and short term. Most obviously, the very act of sponge fishing spread the disease more quickly and effectively than the water current could have alone. A common belief existed that 'sponges contain seeds' and squeezing the sponge 'gurry' into the water released these seeds, starting new growths (Moore, 1910; Anonymous, 1939, p74). This idea was entirely erroneous, but probably contributed to the distribution of virulent microbes among sponge populations. Therefore, the spongers themselves were probably vectors of dispersal, spreading the disease from one isolated bed to another.

Over a longer time-scale, over-fishing very likely increased the concentrations of bacteria and other particles of organic matter – including algal and fungal cells – in the seawater over the sponge beds. The sponges' remarkably efficient filters can remove over 90 per cent of the bacteria from seawater (Reiswig, 1971). Sponges are the only animals on coral reefs that fill this ecological role. The 4 million lb of live sponges taken from the Florida Keys in a typical year would have assimilated more than 70 metric tonnes of organic material annually and taken over 5 tonnes of bacteria from the water – more than 5 million trillion individual bacterial cells (calculations based on data in Reiswig, 1971, 1974). Over the course of the fishery, traditional spongers and divers took more than 1.5 billion lb of living sponges from the northern Caribbean (see Figure 3.7), so that the water during the 1930s was certainly richer in bacteria, fungi, algae and other tiny particles than it had been a century earlier. The community dynamics of microbes in the sea and within sponges are complex; but evidence exists to indicate that at high concentrations, bacteria that are typically benign can become virulent (Miller and Bassler, 2001), and in waters with high levels of organic matter, diseases are more prevalent (Hodges et al, 2005; Kline et al, 2006; Smith et al, 2006). Far from passive free-floating cells, pelagic bacteria are known to attack organic matter and transform biogeochemical cycles (Azam, 1998). It is likely that over-fishing of sponges – the only effective filter feeder in the coral reef ecosystems of the northern Caribbean – fundamentally altered the microbial community in a manner ultimately fatal to their existence (Hobbie et al, 1977; Azam et al, 1983; Moriarty et al, 1985).

Whatever the ultimate cause of the sponge epidemic, the near complete elimination of two genera of marine sponges from the northern Caribbean within one year was not an isolated disaster, but rather linked to a general decline in the integrity of coral reef communities. A measurable trend towards decreased plant and animal biomass and increased bacterial concentrations is a major concern in modern coral reef ecosystems, including those in the Florida Keys (Pandolfi et al, 2005). In pristine reef ecosystems, abundant long-lived corals build habitat for

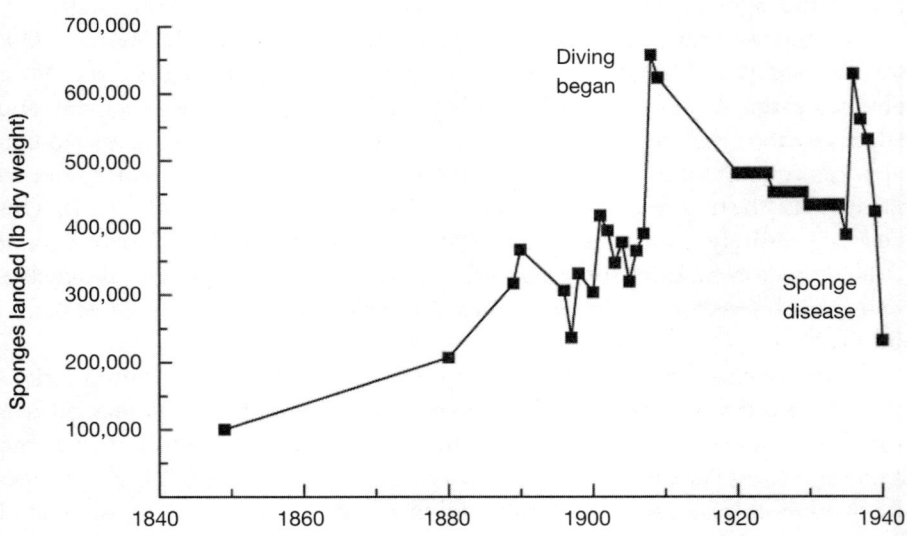

Figure 3.7 *Florida sponge catches, 1839–1940 (lb dry weight)*

Source: data derived from Moore (1910) and Stuart (1948)

large animals, fishes and invertebrates. In contrast, reefs degraded by humans 'are little more than rubble, seaweed and slime' (Pandolfi et al, 2005, p1725). High levels of microbial activity in degraded systems are averse to maintaining biomass in large animals. In ecosystems with dominant microbial communities, fisheries production has even been diminished as a higher proportion of organic matter is locked into the microbial loop (Azam, 1998). It is difficult to demonstrate a link between disease and decreased animal biomass historically; but modern trends of increased disease, diminished water quality and reduced animal biomass certainly suggest that such a connection exists.

In the Florida Keys, the disappearance of commercial sponges was an important component of the unravelling of the coral reef ecosystems, and one with unexpected ecological and social consequences. Along with decreased water quality, the ecological extinction of sponges reduced benthic habitat structure so that in the decade following the 1939 sponge disease, lobsters that are known to live in close association with sponges began to appear in smaller numbers (*KWC*, 1949). Loss of productivity in the lobster fishery is a striking example of how degraded ocean ecosystems become much less valuable to people. Just as the decline of sponges altered the microbial community in a manner ultimately detrimental to sponge health, over-fishing fundamentally altered the reef community in the Florida Keys so that it no longer supports commercially viable fisheries for sponges or lobsters.

Over-exploitation and marine ecological degradation are related to the insatiable demands placed on a resource by external markets – such as those for bath sponges throughout North America during the early 20th century – and to advancements in fishing technology that drive an already stressed resource to commercial and ecological extinction. In south Florida, primitive fishing technology protected sponges far more effectively than either government regulation or the limited conservation ethic of the fishermen could have done; rapid technological advance obliterated these deep-water refuges. Similar scenarios have occurred under the sea and have been particularly harmful to communities of benthic invertebrates. Animals attached to the ocean floor, such as sponges and oysters, are vulnerable in their immobility and irreplaceable in their ecological role. In temperate bays and along tropical coasts, comparatively clumsy, labour-intensive fishing gear protected oysters – filter-feeding animals that provide essential habitat to fish and invertebrates. Fishermen working in sailing vessels along the Atlantic Coast of North America used oyster tongs to grab oysters from their reefs, and free-divers in the Caribbean and tropical Pacific gathered pearl oysters for over three centuries. In both ocean basins, deep-water refuges existed for oysters. During the late 19th and early 20th centuries, however, large-scale dredging and mechanized diving eliminated these refuges and, like sponges, oysters can now be taken wherever they grow. As a consequence, over-fished modern oyster reef ecosystems have been characterized by disease, decreased water quality and loss of architectural structure, as well as the loss of commercial productivity to fishermen.

Particularly in tropical environments with high biological diversity, ecological interactions are complicated and nearly impossible to reconstruct precisely with sparse historic data. By reconstructing the history of the sponge fishermen and sponge community, this chapter has attempted to tell a story of change in both the human and ecological communities. This synthetic approach to historical ecology highlights elements of complex interactions among members of the marine community and between humans and the environments that sustain them. This story of sponge fishing in Florida provides a poignant example of how long-term environmental stress caused by over-fishing lowers biological and commercial productivity, increases the likelihood of punctuated disease events, and takes a harsh toll on people who make their living on the seas.

ACKNOWLEDGEMENTS

Jeffrey Bolster, Marah Hardt, Jeremy Jackson, Matthew McKenzie, Morgan Richie, Daniel Vickers and two anonymous referees provided valuable comments on drafts of this chapter. Thanks to Tom Hambright at the Monroe County Public Library for assistance in locating documents, and to Christina Diaz and Robert Thacker for an underwater introduction to Caribbean sponges.

REFERENCES

Albury, P. (1975) *The Story of the Bahamas*, Macmillan, London
Anonymous (1939) *The Sponge Industry of Florida*, Florida Writers Project, Work Progress Administration, Florida, US
Azam, F. (1998) 'Microbial control of oceanic carbon flux: The plot thickens', *Science*, vol 280, pp694–696
Azam, F., Fenchel, T., Field, J. G., Gray, J. S., Meyer-Reil, L. A. and Thingstad, F. (1983) 'The ecological role of water-column microbes in the sea', *Marine Ecology Progress Series*, vol 10, pp257–263
Barbour, T. (1945) *That Vanished Eden: A Naturalist's Florida*, Little, Brown, and Company, Boston, MA
Blake, H. (1885) 'Letter from the Governor', 8 August 1885, C.O. 23/226, The National Archive, Kew, UK
Brice, J. J. (1897) *The Fish and Fisheries of the Coastal Waters of Florida*, US Commission of Fish and Fisheries, Government Printing Office, Washington, DC
Brodie, J. F. (1994) *Contraception and Abortion in 19th-Century America*, Cornell University Press, Ithaca, NY
Cobb, J. N. (1902) *The Sponge Fishery of Florida in 1900*, US Commission of Fish and Fisheries, Government Printing Office, Washington, DC
Collins, J. W. (1897) *Report on the Discovery and Investigation of Fishing Grounds Made by the Fish Commission Steamer* Albatross *during a Cruise along the Atlantic Coast and*

in the Gulf of Mexico; with Notes on the Gulf Fisheries, US Commission of Fish and Fisheries, Government Printing Office, Washington, DC

Dean, L. (1981) 'Sponger money: It may come back', *Florida Keys Magazine*, Second Quarter, pp10–16

FOR (1931–1939) *The Fish and Oyster Reporter*, Bartow Printing Company, Tampa

FTU (1939) *Florida Times-Union*, newspaper, Jacksonville, FL

Galtsoff, P. S. (1939) *Sponge Mortality in the Key West Area*, Report for US Bureau of Fisheries, Department of the Interior, Washington, DC

Galtsoff, P. S., Brown, H. H., Leslie Smith, C. and Walton-Smith, F. G. (1939) 'Sponge mortality in the Bahamas', *Nature*, Issue 3528, pp807–808

Goode, G. B. (1887) 'Section III: The fishing grounds of North America, with forty-nine charts', in G. B. Goode (ed) *The Fisheries and Fishery Industries of the United States*, US Commission of Fish and Fisheries, Government Printing Office, Washington, DC

Goode, G. B. and Collins, J. W. (1887) *The Fishermen of the United States*, US Commission of Fish and Fisheries, Government Printing Office, Washington, DC

Hall, E. L. and Dietrich, A. M. (2000) 'A brief history of drinking water', *Opflow*, vol 26, no 6, www.wrb.state.ri.us/program_eohistory.html, accessed 24 August 2007

Haynes-Smith, W. F. (1896) 'Letter to Joseph Chamberlain', 7 July 1896, C. O. 23/244, The National Archives, Kew, UK

Hobbie, J. E., Daley, R. J. and Jasper, S. (1977) 'Use of nuclepore filters for counting bacteria by fluorescence miscroscopy', *Applied Environmental Microbiology*, vol 33, no 5, pp1225–1228

Hodges, L. R., Bano, N., Hollitaugh, J. T. and Yager, P. L. (2005) 'Illustrating the importance of particulate organic matter to pelagic microbial abundance and community structure: An Arctic case study', *Aquatic Microbial Ecology*, vol 40, no 3, pp217–227

James, W. (1998) *Holding aloft the Banner of Ethiopia: Caribbean Radicalism in Early 20th Century America*, Verso, London

Johnson, H. (1988) 'Bahamian labor migration to Florida in the late 19th and early 20th centuries', *International Migration Review*, vol 12, no 1, pp84–103

Kline, D. I., Kuntz, N. M., Breitbart, M., Knowlton, N. and Rohwer, F. (2006) 'Role of elevated organic carbon levels and microbial activity in coral mortality', *Marine Ecology Progress Series*, vol 314, pp119–125

KWC (1930–1949) *The Key West Citizen*, newspaper, Key West, FL

KWCHR (1837–1840) Key West Custom House Records, Key West, FL

Lee, K. Y., Lee, J. H. and Lee, H. K. (2001) 'Microbial symbiosis in marine sponges', *The Journal of Microbiology*, vol 39, no 4, pp254–264

Miller, M. B. and Bassler, B. L. (2001) 'Quorum sensing in bacteria', *Annual Review of Microbiology*, vol 55, pp165–199

Moore, H. F. (1910) *The Commercial Sponges and the Sponge Fishery*, US Commission of Fish and Fisheries, Government Printing Office, Washington, DC

Moriarty, D. J. W., Pollard, P. C. and Hunt, W. G. (1985) 'Temporal and spatial variation in bacterial production in the water column over a coral reef', *Marine Biology*, vol 85, pp285–292

Pandolfi, J. M., Bradbury, R. H., Sala, E., Hughes, T. P., Bjorndal, K. A., Cooke, R. G., McArdle, D., McClenachan, L., Newman, M. J. H., Paredes, G., Warner, R. R. and

Jackson, J. B. C. (2003) 'Global trajectories of the long-term decline of coral reef ecosystems', *Science*, vol 301, pp955–958

Pandolfi, J. M., Jackson, J. B. C., Baron, N., Bradbury, R. H., Guzman, H. M., Hughes, T. P., Kappel, C. V., Micheli, F., Ogden, J. C., Possingham, H. P. and Sala, E. (2005) 'Are US coral reefs on the slippery slope to slime?', *Science*, vol 307, pp1725–1726

Powles, L. D. (1888) *The Land of the Pink Pearl or Recollections of Life in the Bahamas*, Sampson, Low, Marston, Searle and Rivington, London

Reiswig, H. M. (1971) 'Particle feeding in natural populations of three marine demosponges', *Biological Bulletin*, vol 141, no 3, pp568–591

Reiswig, H. M. (1974) 'Water transport, respiration, and energetics of three tropical marine sponges', *Journal of Experimental Biology and Ecology*, vol 14, pp231–249

Rolle, T. (1940) 'Sponger money', Recorded in Key West, Florida, 23 January, www.southernmusic.net, accessed 24 August 2007

Schroeder, W. C. (1924) *Fisheries of Key West and the Clam Industry of Southern Florida*, US Commission of Fish and Fisheries, Government Printing Office, Washington, DC

Smith, H. M. (1898) *The Florida Commercial Sponges*, US Commission of Fish and Fisheries, Government Printing Office, Washington, DC

Smith, J. E., Shaw, M., Edwards, R. A., Obura, D., Pantos, D. O., Sala, E., Sandin, S. A., Smriga, S., Hatay, M. and Rohwer, F. L. (2006) 'Indirect effects of algae on coral: Algae-mediated, microbe-induced coral mortality', *Ecology Letters*, vol 9, pp835–845

Stuart, A. H. (1948) *World Trade in Sponges*, Industrial series 2, US Department of Commerce, Washington, DC, pp1–95

Stuart, V. and Klumpp, D. W. (1984) 'Evidence for food-resource partitioning by kelp-bed filter feeders', *Marine Ecology Progress Series*, vol 16, no 2, pp27–37

Sweeting, M. C. (undated) 'Sponger Money Never Done', Traditional folk song, http://faculty.mccfl.edu/Jonesj/Hum2230/SETTLERpoetry.html, accessed 24 August 2007

Tierney, J. Q. (1949) *The Sponge Industry of Florida*, University of Miami, Coral Gables, Florida, US

Townsend, C. H. (1897) *Statistics of the Fisheries of the Gulf States*, US Commission of Fish and Fisheries, Government Printing Office, Washington, DC

Walton-Smith. F. G. (1941) 'Sponge disease in British Honduras, and its transmission by water currents', *Ecology*, vol 22, no 4, pp415–421

4

'Quite the Choicest Protein Dish': The Costs of Consuming Seafood in American Restaurants, 1850–2006

Glenn A. Jones

Major concern over the fate of the world's fisheries has recently been expressed in peer-reviewed scientific journals, newspaper articles, popular books and television documentaries. This concern is fuelled by data showing decreasing standing stocks or biomass of a number of major commercial fish species, reports of the demise of individual fisheries and predictions that the world's commercial fisheries will collapse altogether if current trends do not change (e.g. Worm et al, 2006). These studies have focused largely on the issue of production – that is, the number of fish that live in, and are taken from, the oceans. Emphasis has also been placed on the historical abundance of fish stocks, with researchers endeavouring to establish how many fish there were at a given point in time, the biomass of a particular species prior to the advent of industrial fishing, and the long-term rate of decline of a particular stock. These efforts have entailed analysis of a wide range of archival data to estimate the catch per unit effort (CPUE) of a fishery and from that to estimate the biomass of the target species over a long temporal span (e.g. Rosenberg et al, 2005).

The consumption of fish raises a set of related and equally significant questions: what fish were people eating? Why were they doing so? What were they willing to pay for it? What impacts did their choices have on fish stocks? Yet, this aspect of the global commercial fishing industry has received scant attention from scientists despite the fact that the wealth of Northern Hemisphere regions (specifically the European Union, US and Japan) has created a truly global fish market in which species that have been driven to near commercial extinction in the North Atlantic are now being imported, or replaced, with similar species from other oceans and regions of the world. Examples include Chilean seabass and orange roughy from the Southern Ocean, spiny lobster from Namibia, swordfish from Chile and hoki from New Zealand. Few consumers realize how significantly their buying/eating preferences have influenced the hunt for those species occupying the top trophic levels of the world's marine ecosystems. Pauly and Maclean (2003) state:

> ... we ought as well to be aware of how our eating habits impact [upon] the sea. Sitting back as we read or watch the world news, or drive a car, or even as we walk down the aisle of a supermarket, it is hard to imagine that as individuals we are an integral part of the food web of marine ecosystems.

Although the term 'fishing down marine food webs' (Pauly et al, 1998) succinctly conjures up an image of what is happening to most of the world's fisheries, the reality is far more complex. In a market-driven economy, if there is no consumer demand for a particular species of fish, fishermen will neither target that species nor supply it to the marketplace. This chapter examines the consumption side of the seafood industry from the 1850s to the early 21st century. It presents three case studies which demonstrate that it is the wealthy consumers of the Northern Hemisphere who are 'eating their way down marine food webs'. The analysis deploys data assembled from a hitherto unused source: the printed menus generated by US restaurants over the last 150 years. Although they were commercial documents designed to inform contemporary customers rather than scientists of a later era, bills of fare provide valuable insights into the eating preferences of consumers and the retail prices paid for seafood. Their value lies in the breadth and depth of the information they contain. On the one hand, menus range over a significant segment of the market: it is estimated that during the early 21st century, 60 per cent of the dollar value of the world's fish catch is sold in restaurants. On the other hand, they yield evidence that predates by over a century the data on fish landings and wholesale prices assembled by the United Nations Food and Agriculture Organization (FAO) and the National Oceanic and Atmospheric Administration (NOAA)/National Marine Fisheries Service (NMFS) from the 1950s. In essence, menus offer a novel perspective on the long-term interaction of humans and nature: a view through the eye of the diner.

MENUS AS EVIDENCE

In the US, menus, or bills of fare, debuted in the 1820s and were in widespread use by the 1850s. Although they are by nature ephemeral objects that serve no purpose once the goods they advertise have been consumed, many menus were saved by restaurant patrons, perhaps as a memento of a birthday, anniversary, graduation or other special event, or because the cover art was attractive. People began collecting menus during the 1880s and a number of the early assemblages have found their way into public and institutional archives. In 1915, a 'collection of 76 of the most famous menus' was displayed at the Panama-Pacific Exhibition in San Francisco (Lehner, 1915). A search for historic menus undertaken in 2001 uncovered over 200,000 archived across the US, some of the larger collections being held at the New York Public Library (approximately 35,000 menus), the New York Historical Society (approximately 25,000), the Johnson and Wales Culinary Archives (approximately 50,000) and the Culinary Institute of America (approximately 30,000). A number of smaller collections containing up to a few thousand menus are also housed at repositories such as the Library of Congress, the National Archives and the American Antiquarian Society.

Three distinct types of menu survive in these archival collections. First, there is the hotel bill of fare, which provided a printed list of the food items offered daily in a hotel restaurant; however, since the meal was included in the price of the room none of the items are priced (see Figure 4.1). The banquet menu prepared for annual meetings of clubs and societies, as well as meals to celebrate a milestone (such as the opening of a new building or the retirement of a civic leader, businessman or diplomat), constitutes the second form of menu. Such social events were very popular in the second half of the 19th century, and the menus they generated are the most common type found in the archival collections. Banquet menus have the location and date of the event and a list of food items served, but no prices are given (see Figure 4.2). Third, and most useful for analytical purposes, is the restaurant bill of fare, which normally specifies the price, as well as the identity, of each food item on sale.

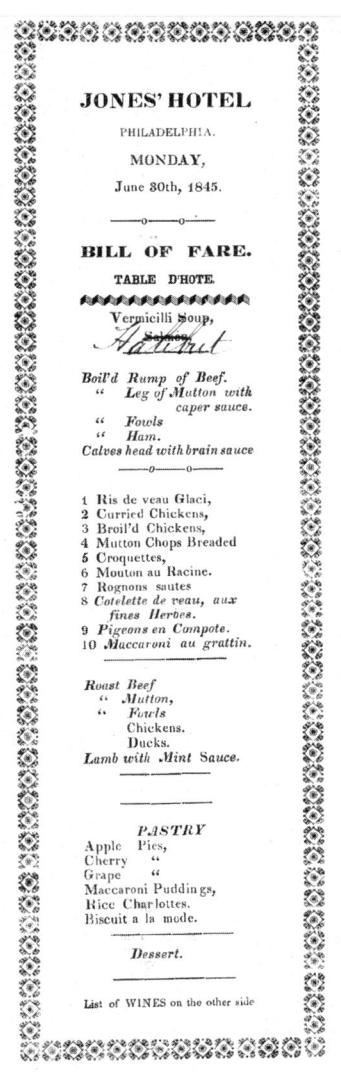

Figure 4.1 *Bill of fare from Jones' Hotel (June 1845)*

Note: Hotel menus typically listed the items offered for the day, but prices were not listed as the meal was included with the price of the room. On this menu, salmon was the fish course that was printed; but that was overwritten by halibut, which was the fish that was served that day.

Source: chapter author's collection

Figure 4.2 *Bill of fare from a banquet celebrating the opening of the New York and Canada railroad (November 1875)*

Note: Banquet dinners became very popular towards the end of the 19th century, and banquet menus are the type most commonly found in archive collections. Although the foods served are listed, no prices are given. For this particular banquet, the fish course consisted of boiled haddock, fried halibut, broiled shad, baked pike and, for the cold dish, one could get lobster salad or pickled oysters.

Source: chapter author's collection

Bill of Fare.

SOUP.
Mock Turtle, or Oefs Empereur.

FISH.
Boiled Haddock, a la Hollandaise,   Fried Halibut,   Broiled Shad,   Baked Pike.

COLD DISHES.
Boned Turkey and Jelly,   Pate of German Hare. a la Strasbourg,
Chicken Salad.   Lobster Salad,   Pickled Oysters,
Smoked Beef Tongue,   Sardines, a l' huile.

BOILED.
Leg of Mutton. Caper Sauce,
Fresh Beef, a la Flammande,   Turkey, with Oyster Sauce.

ROAST.
Saddle of Antelope, with Jelly.   Ham, with Champagne Sauce.   Spring Lamb, with Mint Sauce,
Roast Chicken, with Giblet Sauce,   Partridges, with Currant Jelly.   Plover,
Canvas Back Duck.

ENTREES.
Broiled Pigeon, on Toast.   Chicken's Liver, with Mushrooms.
Fricasse of Turkey Wings, a l' Empereur,   Saiemy of Ducks, with Olives,
Escalop Venison, a la Chasseur,   Petit Pate a la Reine.
Fillet of Beef, with pom Parisienne,   Stewed Rabbits.   Tempale of Maccaroni, a l' Italien.
Queens's Fritters, a la Victoria.

VEGETABLES.
Squash,   Fresh Corn,   Mashed Potatoes,
Green Peas.   Boiled Onions,   Boiled Potatoes,
Stewed Tomatoes,   Lima Beans,   Asparagus,
Cabbage,   Beets.

RELISHES.
Pickled Plums,   Pears,   Peaches,   Tomato Catsup,   Worcestershire Sauce,
Horse Radish,   Chow-Chow,   Pickled Beets,   Mixed Pickles,
French Mustard,   Challenge Sauce,   Ghirkins,   Celery,   Onions.   Olives,
Tomatoes,   English Mustard.   Celery Salt,   Shaker Apple Sauce,
Cranberry Sauce,   Dressing Salad.   Cole-Slaw,
Currant, Cranberry, Crab Apple and Raspberry Jelly.

PASTRY.
English Plum Pudding, with Brandy Sauce,   Queen of Puddings.   Cold Ferina with Cream,
Snowballs,   Syllabubs,   Delmonico Pudding,
Orange Cake,   Almond Cake,   Fruit Cake,   Crullies,   Coffee Cake.   Cream Cake,
Washington Pie,   Lady Fingers,   Cup Cake.   Cookies,   Chocolate Cake,
Gold, Silver and Sponge Cakes.
Vanilla and Lemon Ice Cream,   Lemon Ice,   Wine Jelly,   Lemon Jelly,
Charlotte Russe.   Peach Meringue.
Mince, Apple, Peach, Cranberry, Cherry, Raspberry, Cocoanut, Pumpkin, Custard,
Lemon, Strawberry Pies,
Currant, Cranberry and Crab Apple Jelly Tarts.

DESSERT.
Apples,   Pears,   Oranges,   Malaga Grapes,   Bananas,   Raisins,   Figs,
Almonds,   Filberts,   Pecans,   English Walnuts,   Brazil Nuts,   Peanuts,
Crackers.   Cheese.
GREEN TEA,   OOLONG,   ENG. BREAKFAST,   COFFEE, with Cream,   CHOCOLATE.

PRESERVES.
Strawberry,   Raspberry Jam.   Blackberry Jam,   Peaches,   Pears,   Plums,
Crab Apples,   Cherries,   Pine Apples.

Figure 4.2 *continued*

Figure 4.3 *Bill of fare from Keeler's European Hotel (March 1897)*

Note: This is a good example of a traditional hotel menu in which the food items are offered à la carte and are individually priced. This menu has an extensive four-page list of offerings for 12 March 1897. For example, there are 39 listings for oysters, 15 for clams, 20 fish entrées, as well as lobster, scallops, terrapin, crabs and green turtle. A meal of boiled live lobster was US$0.40 (equivalent to US$8.70 in constant 2004 US dollars).

Source: chapter author's collection

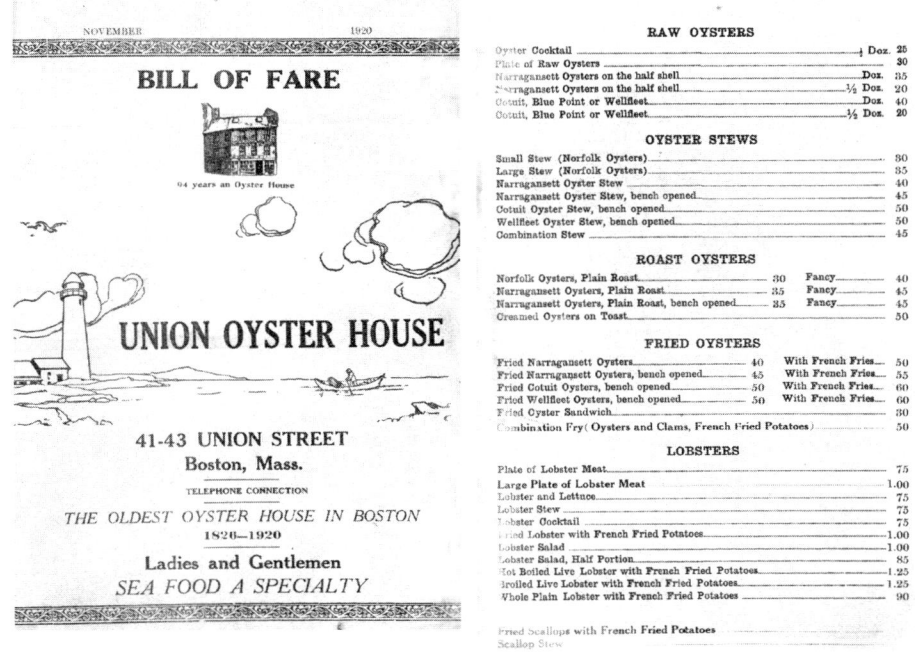

Figure 4.4 *Bill of fare from Union Oyster House (November 1920)*

Note: This is a very traditional format for early 20th-century restaurant menus. It consists of four pages printed on thick card stock and was intended to be used for the entire month of November 1920. A meal of boiled live lobster was US$1.25 (equivalent to US$11.79 in constant 2004 US dollars).

Source: chapter author's collection; view colour version at www.hull.ac.uk/oceanspast

As a source of evidence, the restaurant bill of fare has to be treated with caution for several reasons. First, it has changed in style and substance over the last 150 years. Down to the 1890s, these menus were printed daily on thin paper stock and generally thrown away at the end of each day (see Figure 4.3). After the turn of the century, the bill of fare became more robust, typically being printed on heavier paper, with fancier design elements, so that it could be used for up to a month (see Figure 4.4). By the 1950s, the bill of fare had become more costly for the restaurant to produce and therefore less ephemeral. These menus often have multicolour cover art produced specifically for the restaurant and are printed on robust paper stock intended to last for months. Although these later menus are generally undated, many have an attached dated insert that lists the daily specials, or a printer's code that can be dated, or a contemporary inscription listing the date and type of occasion that prompted the diner to attend that restaurant and save the menu as a memento (see Figure 4.5).

Figure 4.5 *The Fishermen's Grotto menu (July 1957)*

Note: By the mid 20th century, menus had become fancier (and more costly to produce). They rarely had the date printed, so they could be used for several months. Such menus, however, typically had a one-page daily special insert that was often dated, or an inscription from the diner who saved the menu because it commemorated some special occasion. The menu shown here was not dated; but there is contemporary writing on the cover: 'Vacation July 1, 1957'. A meal of fresh Monterey abalone was US$1.75 (equivalent to US$11.75 in constant 2004 US dollars).

Source: chapter author's collection; view colour version at www.hull.ac.uk/oceanspast

Meet your friends at N⁰ 9 Fishermen's Grotto

Owned and operated by

M. Geraldi & Sons

May We Suggest a Cocktail?

OLD FASHIONED	.50
MANHATTAN	.50
SIDE CAR	.60
TOM COLLINS	.50
ALEXANDER	.75

RELISHES AND SEA FOOD APPETIZERS

Ripe Olives 30   Sliced Pickles 15   Celery en Branche 45
Tomato Juice 25   Chilled Clam Juice 35
Jumbo Louisiana Prawns in Shell . . . . 1.20   Half Dozen Eastern Oysters on Half Shell . . . 1.40
Clams on Half Shell . . . (half dozen) . . . .70;   (one dozen) . . . 1.25
Cracked Crab . . . (half) . . . . .85;   (whole) . . . . 1.60   (for two) 1.70

COCKTAILS

Crab Meat Cocktail 60   Shrimp Cocktail 60   Lobster Cocktail 70
Prawn Cocktail 60   Olympia Oyster Cocktail . .70   Eastern Oyster Cocktail . . .70
Combination Sea Food Cocktail . . . Large . . . . 1.10;   Small 60
Crab or Shrimp Cocktail with Mayonnaise or Louis Dressing 65

CHOWDERS

Coney Island Clam Chowder (red) . . . or . . . Boston Clam Chowder (white) . . . or Fresh Clam Broth in Cup 30
(Bread and Butter 15c Extra with Chowder or Cocktails)

TODAY'S SPECIALS

Polpette with Caper Sauce . . . . . . 1.30   Scallops Sauté in Wine a la Newburg . . . 1.60
Rolled Filet of Sole Sauté, with   Filet of English Sole with Fresh Tartar Sauce . 1.40
  Mushroom Sauce . . . . 1.50   Grilled Fresh Sea Bass, Butter Sauce 1.50
Fresh Monterey Abalone, Tartar Sauce . . . 1.75   Grilled Shad Roe with Bacon 1.70*
Fresh Jumbo Frog Legs . . . . 2.20   Grilled Halibut or Salmon, Maitre d'Hotel . 1.50
Grilled Swordfish, Meuniere Sauce . . . . 1.50   Fried Eastern Scallops, Tartar Sauce . . . 1.60
Steamed Clams, Plain or Bordelaise . . . . 1.40   Filet of Barracuda, Grilled, Parsley Butter . 1.40
Spaghetti with Crab Meat, Fishermen's Style . 1.30   Rex Sole or Sand Dabs, Grilled in Butter . 1.50*
Combination Sauté in Butter with Rice . . 1.90   Crab Legs and Shrimps Sauté with Rice,
Genuine Brook Trout, Grilled a la "Grotto" . 1.90   Creole, en Casserole 1.90
Fresh Sea Bass Sauté in Wine, en Casserole . 1.60   Baked Stuffed Turbot in Brown Butter,
Steamed Finan Haddie, Butter Sauce . . . 1.50*   Fishermen's Style 1.80*
Filet of Sole Marguerite . . . 1.80*

OUR FAMOUS SEA FOOD SALADS

CHEF'S SPECIAL — Combination Shell Fish Salad 1.80
California Lobster Salad, Louis Dressing 1.80
All Crab Legs with Louis or Mayonnaise Dressing on the side 1.80
Mixed Green Salad . . .75 . . . With Shrimp or Crab Meat 1.50
Fresh Crab Meat, Louis Dressing . . . 1.50   Combination Crab and Shrimp Salad,
Fresh Local or Northern Shrimps, Louis . 1.50   Louis or Mayonnaise 1.50
Avocado Stuffed with Crab or Shrimp,   Small Prawn Louis 1.30
  Louis or Mayonnaise . . . . . . 1.80   Jumbo Prawns, Louis or Mayonnaise . 1.50
Tomato Stuffed with Crab or Shrimp,   Small Crab or Shrimp Louis 1.30
  Louis or Mayonnaise 1.50

FRESH LOCAL CRAB

Fried Crab Legs, plain . . . . 1.90   Crab Meat a la Newburg en Casserole . 1.90
  a la Bordelaise . . . . 1.90   a la King 1.90
Baked Deviled Crab Meat in Shell . . . . 1.90*   Crab Meat Cakes (2) Italienne 1.60
Crab Meat a la Delmonico . . . . . . 1.90   Baked Crab Meat au Gratin en Casserole . 1.90*
Crab Meat a la Creole or Curried with Rice en Casserole 1.90
Crab Legs Sauté in Wine Sauce . . . . 1.90   Crab Legs a la Mornay 2.00

SHRIMPS AND PRAWNS

Fresh Jumbo Prawns, Fried in Butter . . . 1.65   Shrimps Creole with Rice en Casserole . 1.90
Prawns a la Newburg, en Casserole . . . 1.90   Curried Shrimp with Rice en Casserole . 1.90
  a la King . . . . 1.90   Shrimp Newburg en Casserole . . . 1.90
Shrimp a la King en Casserole 1.80
(Bread and Butter Included with Above Orders)

*Please Allow 15 Minutes to Prepare

Figure 4.5 *continued*

Second, care needs to be exercised in combining data from menus prepared for different social strata. For example, five-star restaurant prices are invariably higher than those of the corner oyster house. Accordingly, to lessen the influence of prices at the top and bottom end of the spectrum, menus from restaurants that are clearly rated 'five star', such as the Waldorf-Astoria, Plaza, Hotel Astor and Delmonico's, are not considered in this study, while restaurants that are obviously local diners and cafeterias are also excluded. Third, descriptions of the food on offer should not be interpreted too literally; for example, it is generally unclear which species of abalone – a shellfish that has experienced serial depletions – was being served. Fourth, bills of fare provide few clues as to portion size, with lobsters described as 'chicken', 'medium', 'large' or 'jumbo', rather than their exact size. Although there are conventions for what these descriptive labels mean, they were not applied consistently by different restaurants. Fifth, the restaurant bill of fare was not produced with archiving in mind, so the temporal and geographic distribution of preserved menus is far from uniform. For some years and in some places, many menus have survived, whereas other times and towns are poorly represented; moreover, the type of restaurant menu that has survived may not accurately reflect the population of seafood-serving restaurants at a particular point in time and space.

Despite these caveats, restaurant menus can be used to shed light on the types of seafood being offered and the prices charged for those items over the long term. A representative sampling of the bills of fare assembled in the above-mentioned collections suggests about 5 per cent (about 10,000 menus) contain the necessary ingredients – date, prices of individual items and restaurant location – for such an analysis. In this chapter, a smaller subset of these menus has been examined to reveal how and why the prices charged by restaurants for three types of seafood have altered over the last 150 years.

ANALYSING THE MENUS

Sumaila (1999) conducted one of the first studies of temporal changes in the cost of seafood. In this analysis, different species were lumped into three categories: invertebrates, large finfish and small finfish. The wholesale cost of these species was obtained from the FAO database and then compared to the Consumer Price Index (CPI). Sumaila (1999) concluded that since the 1950s (as far back as the database extended) seafood prices had risen faster than the inflation rate, suggesting that supply was decreasing due to over-fishing, and/or demand was rising faster than supply. However, a difficulty with this study was that prices of individual groupings of seafood were not normalized to the CPI, and over time the exponential increase in the prices and the CPI makes it increasingly difficult to visually interpret the relationship between the CPI and an individual seafood item (see Figure 4.6).

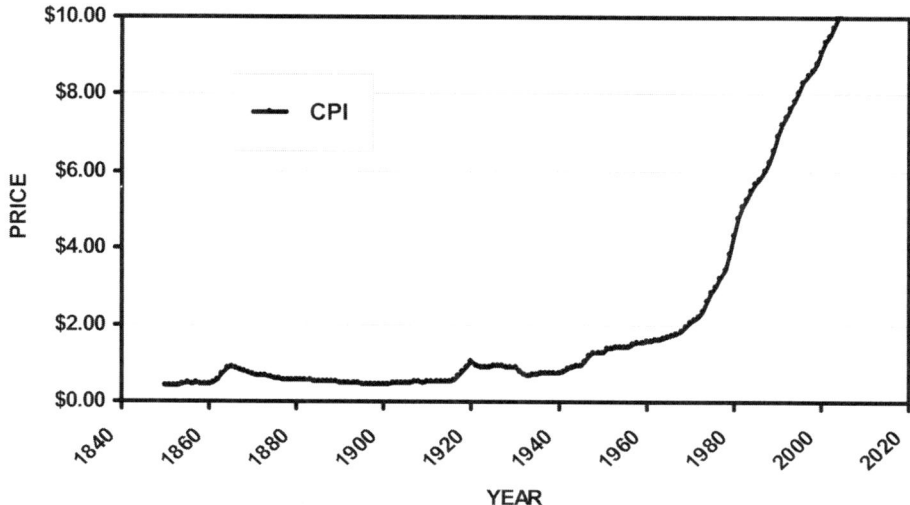

Figure 4.6 *The Consumer Price Index and its pre-1913 equivalent*

Note: Current dollars are non-inflation adjusted. Note the difficulty in identifying any rate of change for the earlier part of the record when looking at current dollars. On this plot, the 1850 value is US$0.43, meaning that the purchasing power of US$0.43 in 1850 is the equivalent of US$10 in 2004. Current dollar prices have increased 23-fold since 1850. The alternative is to use constant dollars where the value is inflation adjusted. In this case, a plot of constant dollars would be a straight line with a value of US$10 for all years from 1850 to 2004 (see Figure 4.7).

Source: data derived from McCusker (2001) and Sahr (2005)

Although the CPI began in 1913, economic historians have reconstructed a price index that goes back as far as the 1660s (McCusker, 2001). This, together with Sahr's (2005) updates, has been used in the current study so that all prices are presented in terms of the value of the US dollar in 2004. The CPI and its pre-1913 equivalent are shown in Figure 4.6 and are referred to as the 'CPI'. Note the exponential increase in the slope of this curve and the difficulty in discerning a rate of change for the earlier part of the curve. It is far more informative to present the data graphically if they have been normalized to the CPI. For example, if the price of a particular item increases over time at the underlying inflation rate (i.e. the CPI), then a plot of that item normalized to the CPI will be a horizontal line as time goes on (see Figure 4.7a). In other words, the price of the particular item has neither increased nor decreased relative to the cost of the representative basket of consumer goods tracked by the CPI. If an item follows the CPI for a period of time and then starts to rise in price at a rate of CPI + 2 per cent, then the plot of the normalized price will rise through time (see Figure 4.7b). Conversely, even if an item rises in price over time but does so more slowly than the underlying

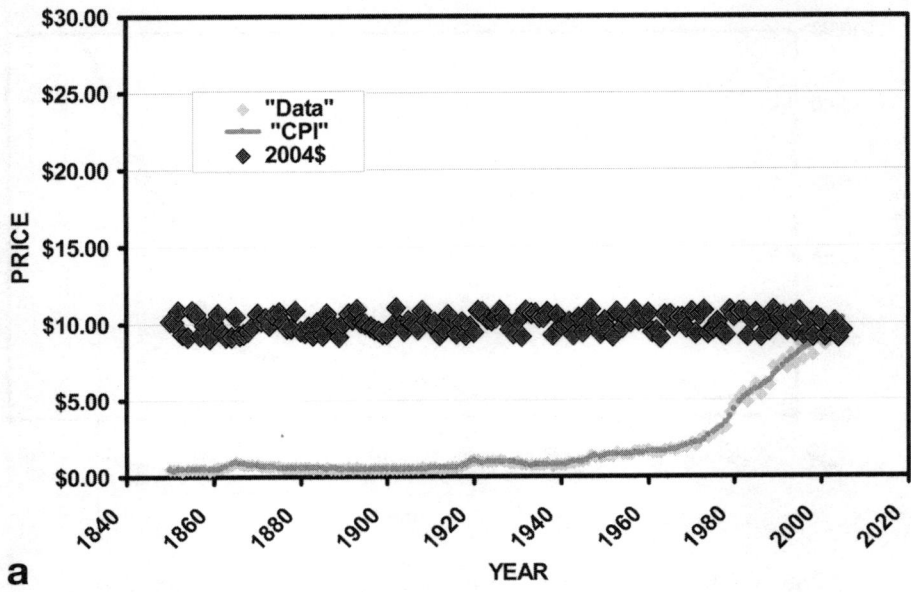

Figure 4.7 *Hypothetical temporal price behaviour for item following: (a) CPI; (b) CPI changing to CPI + 2 per cent after 1950; (c) CPI changing to CPI − 2 per cent after 1950*

Notes: (a) When normalized to 2004 US dollars, this item exhibits a constant price. Even though the current dollar price may be increasing over time, since it is rising at the rate of the CPI, the consumer buying power for that item remains the same in constant dollars.
(b) When looking at the 2004 US dollar normalized prices, one sees a positive (increasing) slope, which indicates that the item is costing the consumer a greater share of his/her constant dollar purchasing power as time goes on.
(c) When looking at the 2004 US dollar normalized prices, one sees a negative (decreasing) slope, which indicates that the item is costing the consumer a lesser share of his/her constant dollar purchasing power as time goes on.

Source: data derived from McCusker (2001) and Sahr (2005); view colour version at www.hull.ac.uk/oceanspast

inflation rate, then the plot of the normalized price of that item will fall through time (see Figure 4.7c).

The importance of using the CPI normalized price is illustrated clearly with the wholesale price of lobster obtained from the NMFS database from 1950 to 2004 (see Figure 4.8). Although a comparison of the CPI and the current dollar price of lobster appears to show that lobster has consistently exceeded the CPI, similar to the analysis of Sumaila (1999), the constant dollar (2004 US dollars) pattern reveals something completely different. The highest rate of increase is from the mid 1950s to mid 1970s, while a decrease in constant dollar pricing is evident from the mid 1970s to the mid 1990s (see Figure 4.8).

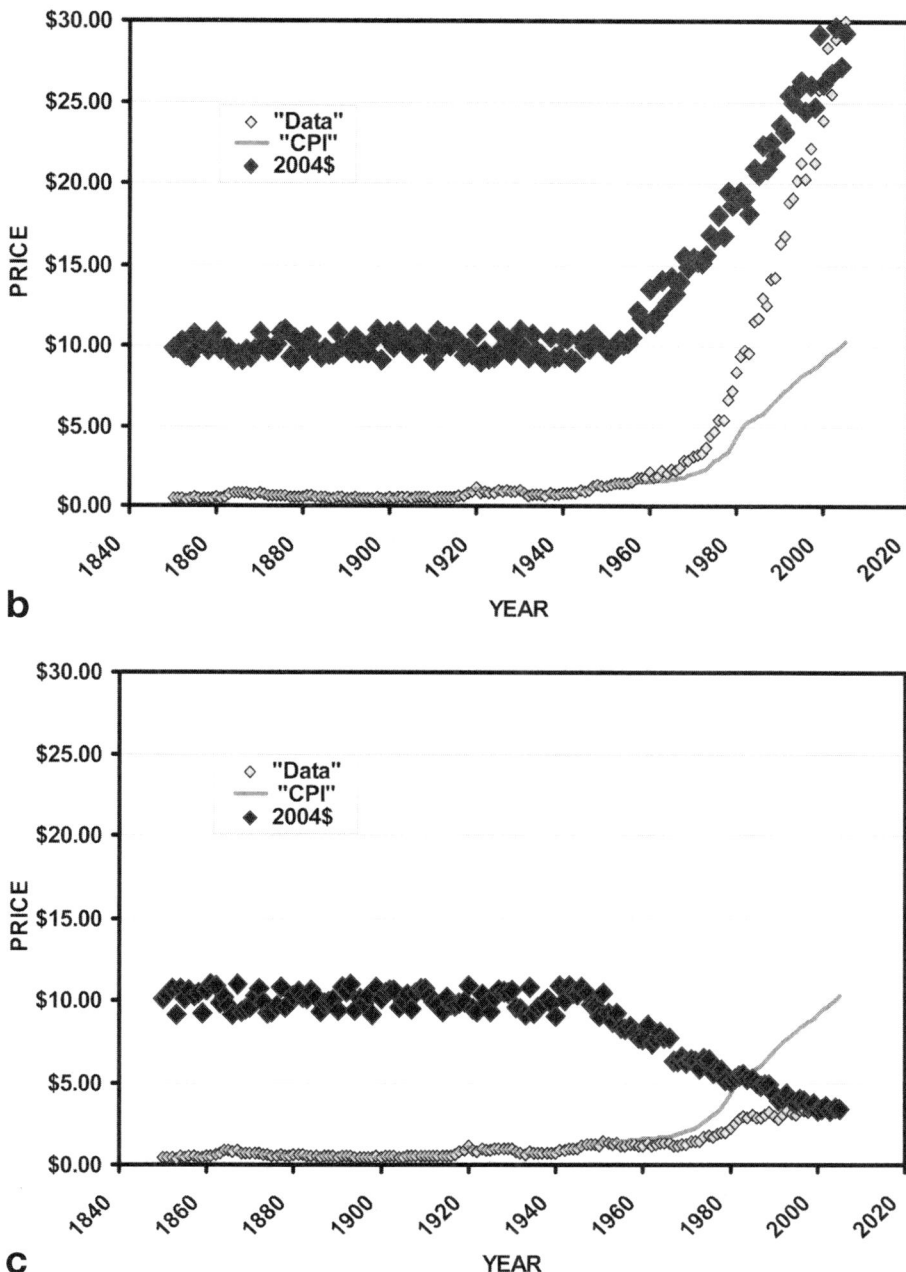

Table 4.1 shows the prices charged for selected seafood items as listed in a menu printed for Bristol's Oyster and Chop House in New York City in January 1900. These prices are normalized to 2004 prices using the CPI. An item costing US$1 in 1900 is valued at US$21.27 in 2004 dollar terms, so a US$0.40 meal of halibut at

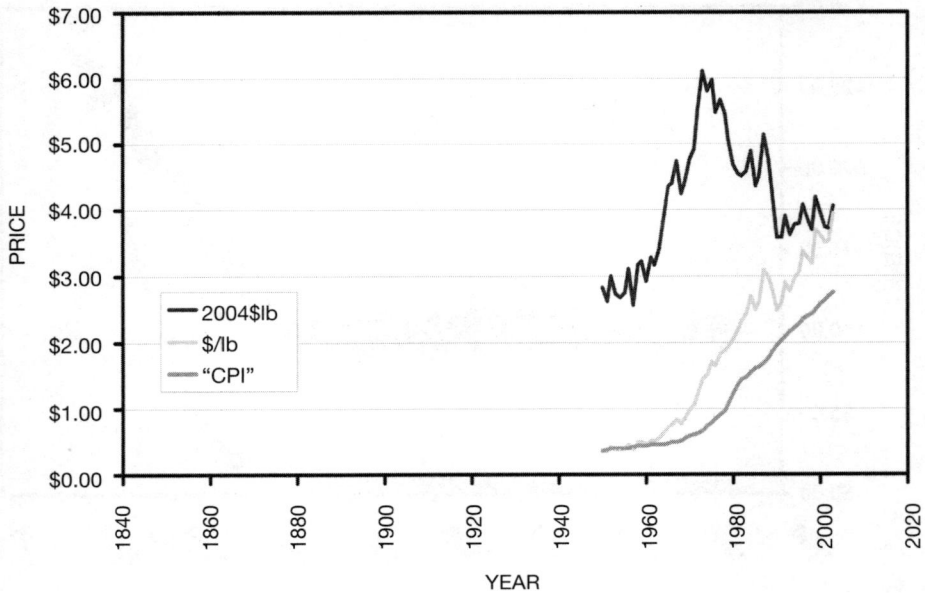

Figure 4.8 *Comparison of the CPI, current dollar and constant dollar wholesale price of live lobster derived from the NOAA fisheries database*

Note: Comparison of the current dollar price of lobster to the CPI suggests that lobster has consistently outpaced the CPI. However, in constant dollar terms, it becomes clear that the price of lobster peaked in 1974 and actually has decreased by 20 per cent (in constant dollar terms) during the last 30 years.

Source: NOAA fisheries database, www.st.nmfs.noaa.gov/st1/commercial/landings/annual_landings.html (accessed 29 August 2007)

the Oyster and Chop House in 1900 would have cost US$8.51 in 2004 if the price of a halibut meal tracked the CPI. Applying this technique to price information garnered from bills of fare dating back to 1850, a database of inflation-adjusted retail seafood prices has been constructed. This facilitates comparative analysis of the price movements of seafood items since the mid 19th century, revealing which items have tracked the CPI, which items have risen (or fallen) faster or slower than the CPI and when changes in trend have occurred.

MENU PRICE CHANGES SINCE CIRCA 1850

To demonstrate the utility of this methodology in relating seafood consumption patterns to stock abundances, case studies of three menu items – the canvas-back

Table 4.1 *Prices charged for selected seafood items by Bristol's Oyster and Chop House, New York City, January 1900*

Fish	Price in 1900 (US cents)	2004 equivalent (US$)
Broiled or fried shad	40	8.51
Codfish steak	30	6.38
Halibut steak	40	8.51
Salmon steak	50	10.63
Bluefish, fried	35	7.45
Bluefish, broiled	35	7.45
Shad roe	50	10.63
Fried smelts	35	7.45
Spanish mackerel	45	9.57
Fresh mackerel	35	7.45
Salt mackerel	30	6.38
Finnan haddie, broiled	35	7.45
Soft shell crabs	50	10.63
Broiled sardines	40	8.51

Notes: Comparing the 1900 prices with the equivalent deflated 2004 US dollar prices indicates that US$1 in 1900 was equivalent in terms of purchasing power to US$21.28 in 2004.

Source: data derived from a menu in the Miss Frank E. Buttolph American Menu Collection, 1851–1930, Item/Page/Plate Number 1900-0107, New York Public Library

duck, the abalone and the American lobster – are presented in this section. Each has been chosen to illustrate a particular point. First, movements in the price of canvas-back duck indicate that increases will exceed inflation if demand remains high while supply is rapidly decreasing. Second, the price of abalone, derived from West Coast (specifically Californian) bills of fare, shows how demand for this slow-growing mollusc altered over time; whereas it was mainly consumed by Chinese and Japanese immigrants before the 1920s, it later became popular with 'mainstream' American diners, a market shift that imposed heavy harvesting pressure on wild stocks of abalone, which have been driven to near commercial extinction in Californian waters. In the third case study, data pertaining to wholesale prices and landings of American lobsters extending back to the early 1880s are used to test the veracity of restaurant menus as a source of information on seafood consumption trends.

Case 1: Canvas-back duck

The canvas-back (*Aythya valisineria*) is the largest of the diving ducks. During the early 21st century, the majority of the population nests on the North American prairies, from Minnesota and the Dakotas in the US to Manitoba and Alberta in

Canada. They migrate, and aggregate, in winter on Chesapeake Bay on the US East Coast, San Francisco Bay on the West Coast and the Mississippi Delta on the Gulf of Mexico. The North American population is approximately 500,000 strong, only 50,000 of which migrate to Chesapeake Bay. Historically, however, the Susquehanna Flats at the head of this bay has seen the largest concentrations of canvas-back ducks due, in large part, to the vast beds of their favourite food, submerged wild celery (*Vallisneria americana*), that grew there. The species-specific name of this duck is derived from the scientific name for wild celery (USFWS, 2007).

Throughout the 18th and 19th centuries, the canvas-back duck was relatively abundant, and a highly regarded delicacy in great demand in major cities, such as Baltimore and Philadelphia, located near the Susquehanna Flats. According to an observer in 1822, the canvas-backs 'sometimes assemble in such multitudes as to cover several acres of the river, and when they rise suddenly, produce a noise resembling thunder'. Several methods of efficiently shooting these ducks were developed, with each hunter commonly killing 200 to 300 ducks per day, largely to supply the dining tables of hotels and restaurants, as the observer related:

> *At public dinners, hotels and particular entertainments, the canvas-backs are universal favourites. They not only grace but dignify the table, and their very name conveys to the imagination of the eager epicure the most comfortable and exhilarating ideas. Hence, on such occasions it has not been uncommon to pay from $1 to $3 a pair for these ducks [US$7.69 to US$23.08 each in 2004 US dollars].* (Anonymous, 1822)

In 1847, it was reported that:

> *... the market of Philadelphia is very plentifully supplied with both canvas-backs and red-heads during the later part of autumn and the winter. The price of the former varies from $1.25 per brace to $3 [US$13.59 to US$32.61 each in 2004 US dollars].... All the hotels of note treat their guests frequently to canvas-backs during the season.... European epicures have long envied the Americans the possession of this splendid bird; but lately the rapid intercourse by steamers between this country has enabled the bon vivant of London and Paris to enjoy the envied American luxury at home.* (Anonymous, 1847)

Despite the apparent plenitude of canvas-backs in the eateries of Philadelphia – and other cities, such as New York, Boston and Chicago, which had all become large consumers of the duck – its numbers were in decline. As a correspondent opined in 1846:

> ... we do not believe that the canvas-back duck, this year, is plentiful; and, indeed, this species has diminished in numbers, in our waters, for some years, owing to the constant warfare that is made against them by the duckers as they are technically called, with their long and large guns. (Anonymous, 1846)

From the 1840s to the passage of the 1918 Migratory Bird Act, numerous articles were published about the changing techniques used to hunt this bird, all designed to supply the demand of the restaurants despite declining numbers. In 1847, for instance, it was suggested that:

> ... the most effectual way of bagging the canvas-back, however, is by shooting it at night. The position of a flock having been previously marked, the sportsman takes to his skiff by moonlight, and by taking advantage of the shadow of the woody bank or cliff, paddles silently to within 15 or 20 yards of a flock of 1000, among whom he makes great slaughter. (Anonymous, 1847)

Half a century later, Fay (1896) outlined the impact of technological change on the hapless prey:

> Once within our borders, the unfortunate fowl discover that every feeding ground at which they may be tempted to alight is garrisoned by human foes, equipped with every deadly device to lure and destroy. Nor is there any rest for them, day or night. With the advent of modern weapons has come the cold-storage system, by which all flesh may be preserved for an indefinite period in frozen condition. Previously, wild fowl were measurably free from molestation in the extreme southern states during the winter months. The refrigeration process has changed all that. With the introduction of this device, the former respite which was granted them has ceased, and their killing goes on mercilessly at the south during the winter as in the northern states at other seasons of the year. Nor is that all. When the fowl start on their northward flight in the spring, they are harassed with the same persistency as during their progress to the south in the autumn. (Fay, 1896)

As the 19th century came to a close it was becoming increasingly evident that drastic measures were needed to prevent the extinction of the migratory waterfowl in North America. Several states passed legislation banning the export of wildfowl and game beyond their borders, and New York passed legislation prohibiting the shooting, or possession, of migratory waterfowl in the spring. As one reporter explained:

> *New York is the great market of the Atlantic seaboard for game of all sorts, including wild fowl. If market gunners find that wild fowl cannot be sold in New York the inducement to kill them almost ceases to exist. Recently, in some of the southern states, and especially along Currituck Sound, the price of canvasbacks fell to 75 cents a pair [US$7.65 each in 2004 US dollars] – a price lower than they have brought for many years, the usual price to gunners being $2.50 or $2 per pair [US$25.50 to US$20.41 each in 2004 US dollars]. The local gunners could not understand what such a fall in price meant; but since they could not get more than this for their birds, many of them have given up shooting, declaring that rather than kill canvas-backs at such a price they would let them go, to return to their northern homes and breed to bring back others next autumn.* (Anonymous, 1904)

Despite the efforts of individual states, the number of illegally shot and transported migratory birds remained high. During the early 20th century – to cite but two out of many examples – some 50,000 illegally shot and/or transported migratory ducks, reed birds and songbirds were discovered at the storage depot of the Arctic Freezing Company ready for illegal sale to the restaurants of New York City, while a further 2000 canvas-back and red-head ducks illegally shot in Minnesota were confiscated *en route* to Chicago's dining rooms (Cristadoro, 1903). The populations of migratory waterfowl continued to fall despite the efforts of individual states to try to control the harvest and export of their game and wildfowl. With a patchwork of legislation it had proven too easy for the gunners to kill their prey in a state where shooting was legal, then illegally transport the catch to another state, before conveying it to one of the major cities. With the passage of the 1918 Migratory Bird Act, all states came under the same rules and it became illegal to transport across state lines or to sell migratory birds. Henceforth, the large-scale commercial shooting ceased and restaurants were unable to serve canvas-backs and other protected species of bird.

Despite its declining population, the canvas-back duck remained in high demand and continued to be listed on restaurant bills of fare until 1918. Menus indicate that the inflation-adjusted price of this popular waterfowl increased continuously and dramatically from approximately US$20 (2004 US dollars) during the 1860s to US$100 (2004 US dollars) by the 1910s (see Figure 4.9), a fivefold rise that exceeded the inflation rate by 3.5 per cent for each year in the 1860 to 1910 period. Few items can be found on 21st century menus that fetch such a price (i.e. US$100 per meal), and having consistently exceeded the underlying inflation rate by so much and for so long. Despite less expensive alternatives, such as the red-headed duck, demand for canvas-back duck remained high, even as supplies of the bird dwindled and its price rose. Indeed, it was precisely because of the increase in price that the canvas-back remained on restaurant menus, for ordering this delicacy not only became an act of conspicuous consumption, but

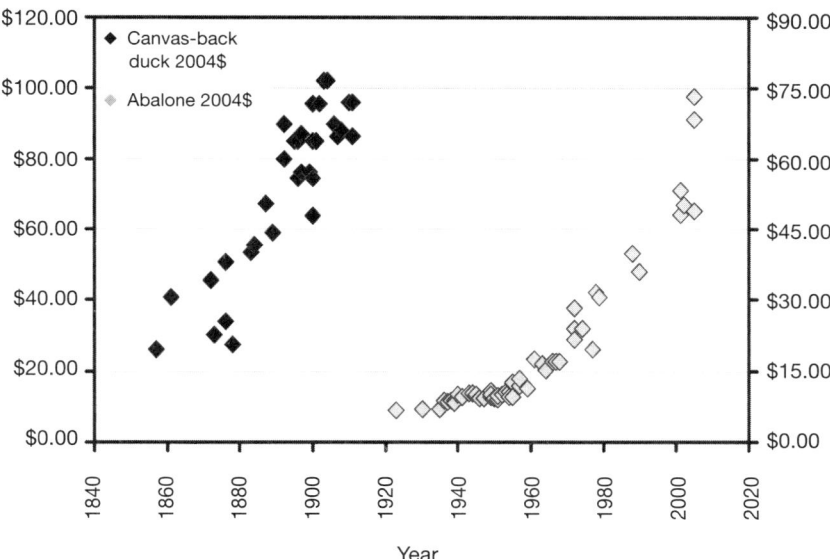

Figure 4.9 *Inflation-adjusted price of canvas-back duck, 1860–1910, and abalone, 1920–2004*

Notes: (left) The inflation-adjusted price increased fivefold from the 1860s until 1918 when commercial sale of the canvas-back duck was prohibited by passage of the Migratory Bird Act. Demand for this species remained very high throughout the 19th century even while supplies dwindled due to excessive commercial hunting. As supplies decreased and demand remained high, the constant dollar price increased. During a near 50-year period the constant dollar price of a canvas-back duck meal increased at an average annual rate of 3.5 per cent higher than the CPI. *(right)* Abalone became popular with the public after the Panama-Pacific Exhibition of 1915. Demand has remained at a high level for the last 80 or so years despite severe population declines and commercial harvesting bans. As supplies decreased and demand remained high, the constant dollar price has increased tenfold. Over the last 80+ years, the constant dollar price of an abalone meal increased at an annual rate that was 4 per cent higher than the CPI in each of those 80 years.

Source: data derived from various menus (see text)

also a symbol of the diner's status and affluence. It was therefore the passage of the 1918 Migratory Bird Act, rather than market forces, which prevented this and other species of migratory waterfowl from literally being eaten into extinction. Such consumption patterns are still evident, with wealthy European, North American and Japanese diners ordering increasingly expensive dishes made from dwindling stocks of top predators such as blue-fin tuna and swordfish, and slow-growing species such as Chilean seabass and orange roughy.

Case 2: Abalone

Several species of abalone are harvested for human consumption worldwide, and all are from the genus *Haliotis*. Traditionally, the most commonly harvested species in California were the red (*H. rufescens*), which is the world's largest abalone, pink (*H. corrugata*), black (*H. chracherodii*) and green (*H. fulgens*) abalone. Since 1997, all commercial harvesting of abalone in California has been banned, and recreational harvesting is limited to the taking of red abalone from the waters north of San Francisco during seven months of the year. While free-diving (no SCUBA) is the only permissible harvesting technique, only four abalone of a minimum diameter of 7in (which yields about 1 lb of meat) can be collected in a day, with no more than 24 taken in a year. It is illegal to sell any recreationally caught abalone. Red abalone is grown in aquaculture farms, each individual taking four to five years to reach a marketable size of 3.5in to 4.5in in diameter and selling for approximately US$115 per pound. A large illegal market for poached abalone originating from the waters of northern California has developed since 1997, with the black market price for a single 7-inch diameter abalone ranging from US$80 to $100. Restaurants serving legally farmed or poached abalone charge approximately US$75 per meal, which comprises approximately 4 ounces of meat.

Although abalone had been harvested and consumed by Chinese immigrants as early as the 1860s, it was not until the 1920s that it appeared on the menus of non-ethnic restaurants. Contemporary popular literature of the time reveals why this was the case. In 1872, it was reported that 'there are at present in the business of taking and drying abalones on the coast of Lower California about 25 Chinamen, who manage to procure about 6 tonnes per month'. The correspondent went on to describe how 'after a thorough drying it is nicely packed in strong sacks and shipped to San Diego, to be reshipped from here to the Chinese merchants in San Francisco. The meat commands in that city from 5 to 6 cents per pound [US$0.75 to $0.90 in 2004 US dollars]' (Anonymous, 1872). At US$115 per pound, the price of farm-grown abalone in 2007 was more than 100 times the inflation-adjusted price of 1872! Further detail was provided by an observer in 1881:

> In San Francisco and the coast towns, abalone is rarely eaten except by Chinese, who are the only ones who gather it. A simple process of salting and drying is all that is necessary for its preservation, in which shape it is sent to China. In order to get 1 tonne of meat, about 6 tonnes of living animals must be collected; but there is no telling how many individuals this represents. After being cured, abalone meat is worth from 5 to 10 cents a pound [US$0.88 to $1.75 in 2004 US dollars], and the value of the crop which reached San Francisco last year approached US$40,000 distributed among some hundreds of men. The coast is now so stripped of the haliotis that the Chinese fishermen are compelled to resort to unfrequented islands, transportation to which is afforded

> them by American capitalists, who take their pay in shells, while the Chinese retain the meat. (Anonymous, 1881)

During the 1890s, there was still no evidence that abalone had found a market outside the ethnic Chinese community. According to Caton (1891), abalone was:

> ... edible when properly prepared, but are not in common use among the Americans, who do not understand the mode of preparing them. ... In some parts of the coast they abound in great numbers and several hundred pounds of them are frequently captured at one tide. These fish have commercial value now. They are taken in great numbers by Chinamen, who prepare them by cutting them into thin sheets when they are dried, and packed and exported to China, where they are in good demand at about 25 cents a pound [US$5 in 2004 US dollars].

In 1900, laws were passed prohibiting the harvest of abalone from the intertidal zone, which effectively put the Chinese fishermen out of business. This led to Japanese fishermen taking over by using diving equipment to harvest abalone in the subtidal environment. Ware commented in 1903 that the Japanese fishermen:

> ... have a cheap diving suit, with a galvanized helmet, and the air is supplied by a two-cylinder pump. The diver – only one dares to go down – takes to the water at 9, comes up at 12, is down again from 1 until 4. He takes down with him a net-like basket, which he sends up in about 20 minutes, another being sent down while it is emptied. This holds about all that four men can lift into the boat. (Ware, 1903)

The average catch was 2300 abalone per day. After being returned to the fishing camp, the abalone were processed, dried, packed into sacks weighing 200 pounds and sold to Chinese merchants in San Francisco. The price received was about US$5 to $6 per 100 pounds (US$1.02 to $1.22 per pound in 2004 US dollars).

The first indications that abalone was making its way into non-ethnic markets occurred in 1905, when an article in the *Scientific American* reported:

> Though the flesh of the abalone is a nutritious and wholesome article of food, highly esteemed by the Chinese and Japanese, few people in the United States know anything about the abalone, except that it is a large shell with a bright, pearly interior. ... A few years ago some Americans, whose attention had been drawn to the large quantities of dried abalone exported to China and Japan, attempted to discover a process whereby the flesh of the abalone could be rendered soft and palatable. ... After many experiments, a San Franciscan named J. W.

Figure 4.10 *Postcard of Abalone drying on the shore*

Note: Although abalone did not become popular with the mainstream public before 1915, it was highly sought after by the Japanese and Chinese as early as the 1870s. The constant dollar wholesale price of abalone has increased 100 times over the last 130 years. This is equivalent to an annual constant dollar rate of increase of 3.75 per cent higher than the CPI for each of those 130 years.

Source: Inkersley (1905)

> Gayetty discovered a process whereby the flesh of the abalone is rendered soft and succulent like that of an oyster. He is now the president of a company operating a cannery fully equipped with every requisite for the preparation of abalones as food.... A great quantity of canned abalone is sent to China and Japan where it is a favourite article of food. Dried abalone fetch from 11 to 15 cents per pound [US$2.24 to $3.06 in 2004 US dollars]. The leading hotels and restaurants of San Francisco now have abalone chowder as a regular item of their bills of fare. (Inkersley, 1905)

The extent of the abalone fishery in the 20th century is indicated in Figure 4.10, which shows the large stretches of the shoreline dedicated to the drying of abalone by Chinese and Japanese fishermen. The process of drying would take 30 to 60 days from harvest to sacking for sale in San Francisco, leaving the abalone with the texture of dried leather. After Gayetty began canning large quantities of abalone, the time from harvest to market was drastically reduced and large quantities could be processed, as Figure 4.11 implies.

Figure 4.11 *Postcard of abalone shells lying outside a processing plant in Santa Barbara, California*

Note: Being a slow-growing mollusc, this type of commercial harvesting was little more than strip-mining of the resource. Accordingly, the constant dollar price of abalone meat increased dramatically as the supply dwindled. The scale of this commercial operation is in stark contrast to the heavily depleted populations and commercial harvesting ban in effect today.

Source: chapter author's collection; view colour version at www.hull.ac.uk/oceanspast

By 1913, as Edholm (1913) revealed, abalone was becoming increasingly popular beyond the ethnic communities:

> *A delicacy from the sea, which Americans on the Pacific Coast are just beginning to appreciate, is the abalone, a mollusc which grows to a size of 10 inches or more in diameter.... But the Chinese and Japanese have always regarded the mollusc itself as a great delicacy, and in those countries the price of 90 cents a pound is paid for the dried meat [US$17.30 in 2004 US dollars].... A few restaurants on the Pacific Coast have undertaken to serve this shellfish to American patrons; but the method of preparing it is far less complicated [than employed by the Japanese and Chinese], and the results are so much more appetizing that the public may become educated up to placing the abalone on the menu within a few years. An excellent way of serving it is to make it into chowder, just as clams are prepared, while another way of serving it is to slice it very thin, pound until tender, and fry like a steak. It is*

> understood that the fresh abalone is used for this purpose, and not the dried product. Served in either style, it is a most delicious addition to our bill of fare, besides being as wholesome as any other shellfish. When the American public demands this new food, it will greatly increase the industry on the coast, and take it to a great extent out of the hands of the Japanese and Chinese.

Abalone became much more widely known as a result of the 1915 Panama-Pacific Exposition held in San Francisco. California exhibited 365 varieties of local sea fishes, with examples of how best to cook them, as well as building a 50 × 100 foot (15m × 30m) seafood kitchen and a dining hall capable of seating 1300 and 'arranged to give the illusion of a cove on the sea shore ... with dining tables and appliances for steaming and otherwise preparing all sorts of shellfish indigenous to Californian waters, including the succulent aristocrat peculiar to our coasts, the univalve abalone' (Anonymous, 1914). Moreover, as part of an effort to bring California seafood to the attention of the public, the organizers published a menu and recipe book (Lehner, 1915). In effect, the exposition was a defining moment in the ascent of the abalone to the rank of seafood 'aristocrat', and from this point abalone chowder and abalone steaks became commonplace on restaurant menus. With abalone's growing 'mainstream' popularity came the passage of a variety of local and state-wide laws and regulations limiting the recreational and commercial harvest of abalone, by virtue of which size limits were established, methods of harvesting were controlled and exports from California banned. Richey (1925) described this transition:

> Long before the Californian, busy with his struggle with the west, knew or cared to protect the abalone, the Japanese established tiny fishing camps and took these molluscs from the water in such enormous quantities there was danger of their extermination. These fishermen dried the abalone and shipped the product to China. ... Now, the laws of California amply protect the abalone. There is a size limit law which allows the abalone to pass through two or three breeding seasons before it reaches the required size. Another law prohibits shipping abalones from the state either fresh or canned and the drying of abalones is now taboo here.... [Abalone steak is] quite the choicest protein dish that California has to offer.... To get this on the market in prime condition requires quick, efficient work by several workers. At the wharf in Monterey, they are prepared on a large scale.

The surviving evidence indicates that abalone chowder and abalone steak first appeared on restaurant bills of fare in 1916 and 1923, respectively. During the latter year, a meal of fried abalone steak cost US$0.60 (equivalent to US$6.50 in 2004 US dollars). From then until the mid 1950s, the price of an abalone meal exceeded the CPI by approximately 1.5 per cent for each of those 30 years; but from the mid 1950s, the inflation-adjusted price began to rise much faster (see Figure 4.9). Whereas in 1954, the price of a meal was US$1.50 (US$10.56 in 2004 US dollars), it stood at US$75 in 2004. This was an increase of 7.5 times, equivalent to an annual rise of 4 per cent above the underlying inflation rate for each of those 50 years, which resembled the price trajectory observed for the canvas-back duck. During the 1970s, there were dramatic reductions in the harvest of abalone, especially black abalone, and periodic closures of the commercial and recreational fisheries. By 1997, the commercial abalone fishery had been closed permanently, and in December 2006 the Centre for Biological Diversity petitioned the federal government to list the black abalone as an endangered species (CBD, 2006).

Although there are no restaurant menu prices for abalone steak prior to 1923, the inflation-adjusted price of abalone meat has increased 100-fold since 1872. This is equivalent to an annual increase of 3.75 per cent above the underlying inflation rate for each of the 130 years from approximately 1870 to 2000, and is comparable to the 4 per cent annual increase in the inflation-adjusted restaurant menu prices of abalone steak since the 1950s. Here is a seafood item that was ignored by the American consumer until the early 20th century, but has since become a great delicacy commanding a very high price. The abalone fishery is following the path trodden by the canvas-back duck market some 100 years earlier – that is, a high rate of inflation-adjusted price increase over time, population decline, strong consumer demand, growth in illegal trade and, ultimately, closure and sale prohibitions. If lessons are learned from history, then nothing short of a total ban on recreational harvesting – and the elimination of the illegal trade in wild-caught individuals – will save the California abalone from extinction.

Case 3: American lobster

Comparatively full records exist with regard to landings and prices of the American lobster (*Homerus americanus*). These are documented in the NOAA/NMFS dataset from 1950 to the present, and in the State of Maine (the major source of lobster in the US) dataset from the 1880s to the present. Covering some 125 years, these data facilitate comparison with restaurant prices to judge if the latter do, in fact, track wholesale prices.

Before the 1870s, live boiled (or broiled) lobster was rarely offered on restaurant menus. Instead, the most common dishes were lobster salad and lobster meat, largely because individuals of 10lb to 20lb (4.5kg to 9kg) were commonly harvested up to that point, the majority being processed in canneries located along the coast of Maine until 1895 and Canada thereafter (Woodard, 2004). The canneries closed in Maine due to the decreasing size of the average lobster and the increasingly strict regulations imposed on harvesting the species. Moreover, it was more cost-effective to ship canned lobster meat than to convey live lobsters across the country. A fish

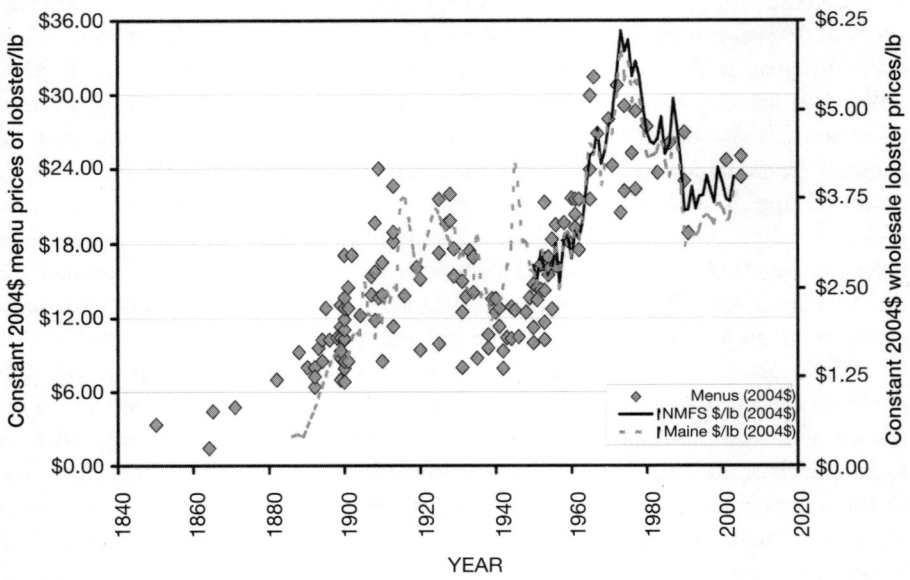

Figure 4.12 *Price series, American lobster, 1885–2004*

Note: The price of a lobster dinner rose significantly from the 1880s to 1920s and from the 1950s to the 1980s, exhibiting annual constant dollar increases that exceeded the CPI by up to 6.5 per cent during each of those nominal 30-year periods. A lobster dinner in 2007 costs the same as it would have done in the late 1920s when prices are adjusted for inflation. The restaurant prices closely follow the wholesale pricing. See text for further explanations.

Source: data derived from: various menus (see text); NOAA fisheries database, www.st.nmfs.noaa.gov/st1/commercial/landings/annual_landings.html (accessed 29 August 2007); and State of Maine, Historical Summary of Maine Lobster Fishery, www.maine.gov/dmr/rm/lobster/lobdata.htm (accessed 29 August 2007); view colour version at www.hull.ac.uk/oceanspast

dealer's price list from 1881 indicates that 12 one-pound cans of lobster meat cost US$1.40 (or US$0.12 per pound of lobster meat), equating to a per pound price of US$2.26 in 2004 US dollars. During the early 21st century, lobster meat is sold frozen for US$35 to $50 per pound, a 15- to 20-fold increase over the underlying inflation rate from 1881 to 2006, which is equivalent to an annual increase of 2 per cent above the underlying inflation rate for each of the last 125 years. The price of a live-boiled lobster meal increased about ten times more than the underlying inflation rate during this same period. However, the price of lobster did not increase steadily during this period. Figure 4.11 shows the inflation-adjusted prices for restaurant offerings of live-boiled (or broiled) lobster, the wholesale price of lobster from the state of Maine database, and the price per pound of live lobster from the NOAA database. Each of these price series has been normalized to the price per pound in 2004 US dollars. It is readily apparent that there is a strong correlation between the retail (restaurant menu) prices and the wholesale prices, with the former being approximately 5.75 times the latter, reflecting the additional cost of items such as rent, labour, profit and accessories served with the lobster.

The state of Maine data suggest that the inflation-adjusted wholesale price of lobster increased from US$0.50 in 1885 to $3.50 in 1915. This sevenfold increase over 30 years is equivalent to an annual increase that exceeded the underlying inflation rate by an average of 6.5 per cent in each of those 30 years. The inflation-adjusted price of lobster listed in restaurant menus increased from approximately US$3 in 1860 to $21 in 1910, an annual increase of 4 per cent above the underlying inflation rate for each of these 50 years. This period exhibits some of the largest inflation-adjusted prices seen for any seafood item, which caused some comment in the popular press:

> *With lobsters at 30 cents a pound they are quite beyond my reach; and all I can do is dream of the time when they sold for 5, and we had all the lobsters we could eat.... When lobsters went up to 6 cents a pound we still bought them just the same; but when they went to 8 we began to fear that the time might come when we should have to do without lobsters; and now with lobsters as 30 cents a pound, that time has long since come. We have no lobsters now save only in our dreams.*
> (Anonymous, 1905a)

In a similar vein, a cartoon published in the satirical magazine *Puck* depicted an auctioneer standing on a table of an expensive restaurant, while the patrons dressed in tuxedos and gowns bid furiously for a lobster being held up by the chef (Anonymous, 1905b). Entitled 'The Passing of the Lobster', and captioned 'Auctioneer of the future – What am I offered for this whole portion of genuine *Homerus americanus*?', the cartoon marked the lobster's passage from food of the peasant to repast of the wealthy.

In 1925, the wholesale price of lobster was US$3.50 (in 2004 US dollars); but it fell throughout the Great Depression of the 1930s to reach US$2 (in 2004 US dollars) in 1939. This decrease in inflation-adjusted dollars was 4 per cent less than the underlying deflation of prices on the CPI during each of these 14 years. A similar decrease was evident in the menu prices of lobsters (see Figure 4.12). Again, history was repeating itself. As the economy plunged further into depression fewer individuals could afford to go to restaurants, and even fewer could afford the price of lobster. This collapse in the market led to fewer lobsters being caught as harvesting effort would likely yield unsold and unprofitable seafood. A similar price collapse had occurred during the early 20th century when New York prohibited the sale of canvas-back duck in the spring, its price collapsed and the commercial gunners stopped shooting.

During World War II, the restaurant price of lobster remained constant because the Office of Price Administration (OPA) fixed the price of various consumer goods to stabilize the economy. An OPA statement was printed on all restaurant menus from 1943 to early 1946 asserting that prices would not exceed those charged during the week of 3 to 10 April 1943. However, lobster was considered a luxury good and the wholesale price was not regulated, so it was canned and sent to the troops as an additional source of protein. These actions caused wholesale prices to rise dramatically while the retail prices charged by restaurants remained constant. After 1945, the price of lobster in restaurants once again exceeded the underlying inflation rate, rising from US$10 (2004 US dollars) in 1946 to $30 (2004 US dollars) per meal in 1975. This threefold increase is equivalent to an annual increase of 3.8 per cent above the underlying inflation rate for each of the 30 years from 1946 to 1975.

During the 1980s, landings of lobster in Maine began to increase dramatically, and there was a concomitant decrease in price. From 1975 to 2002, in terms of 2004 US dollars, the wholesale price of lobster fell from US$5.67 to $3.71 per pound in 2002, whereas the restaurant price of a meal fell from US$29 in 1974 to $23.40 in 2004. This is equivalent to a 1.5 per cent (wholesale) and 1 per cent (restaurant) annual decrease in inflation-adjusted prices during these 30-year periods. In fact, the price of lobster in 2004 was the same in real terms as that being paid during the mid 1920s. Despite this apparent stability, a number of price fluctuations occurred during this 80-year period, which followed in the wake of the inflationary surge of the 1880 to 1920 period, when the broiled lobster meal became widely popular, the image of lobsters began to appear on restaurant menu covers and lobster was no longer the meal of the poor.

Conclusion

The three case studies presented in this chapter demonstrate the utility of restaurant menus in charting seafood prices and consumer preferences since the mid 19th century. In the case of the canvas-back duck, prices exceeded the underlying inflation rate by an average of 3.5 per cent per year from the 1860s to 1918, when the trade in migratory waterfowl was prohibited. Whereas a similar trajectory is evident with regard to the price of abalone since the 1920s, the inflation-adjusted prices of lobster increased some 6.5 per cent above the underlying inflation rate from the 1880s to the 1920s. Moreover, when landings of lobster increased dramatically during the late 20th century, its price in 2004 US dollar terms was some 20 per cent lower than the level attained during the mid 1970s, and equivalent to the price paid by diners in the mid 1920s. Such price movements were conditioned by the forces of demand and supply, with the growth of the former leading to the depletion of the populations that supplied the market. This, in turn, affected patterns of demand as consumers chose either to purchase items that were increasingly scarce and therefore costly – because of the social status afforded by the purchase of such 'emerging' luxuries – or to buy cheaper alternatives.

Taken together, the case studies indicate that retail prices in seafood outlets exceeded the economy-wide rate of inflation from at least the 1850s. This infers that commercial fishing was exerting considerable pressure on marine animal populations long before the 1950s, which has often been cast as the decade that witnessed the advent of industrial-scale fishing. It also emphasizes the role of consumption in shaping the long-term abundance, diversity and distribution of stocks. Ultimately, it is consumers who determine how much they are willing to pay for a particular type of seafood, and as species become scarcer and prices increase in real dollar terms, it will be consumer demand for cheaper alternatives that will condition which species become fished harder to supply that demand.

References

Anonymous (1822) 'American natural history: The canvas-back duck, *Anas valisineria*', *The Port Folio*, vol 1, pp18–21

Anonymous (1846) 'Sporting on the Potomac', *Spirit of the Times*, vol 16, p497

Anonymous (1847) 'Game-birds of America No. V: Canvas-backed duck', *Graham's American Monthly Magazine of Literature, Art, and Fashion*, vol 30, pp162–164

Anonymous (1872) 'Salt water beefsteak', *The Youth's Companion*, vol 45, p6

Anonymous (1881) 'Abalone jewellery and its source', *Harper's Bazaar*, vol 14, pp38–39

Anonymous (1901) 'Why shooting is poor', *Forest and Stream*, vol 57, p202

Anonymous (1904) 'New York spring shooting', *Forest and Stream*, vol 62, p2

Anonymous (1905a) 'Lobsters at thirty cents!', *Massachusetts Ploughman and New England Journal of Agriculture*, vol 64, p6

Anonymous (1905b) 'The passing of the lobster', cartoon, *Puck Magazine*, vol 57, p16

Anonymous (1914) 'Of interest to fishermen at the Panama-Pacific International Exposition', *Field and Stream*, vol 83, p755

Caton, J. D. (1891) 'Abalone and squid', *Forest and Stream*, vol 36, pp474–475

CBD (Centre for Biological Diversity) (2006) 'Petition to list the black abalone (*Haliotis chracherodii*) as threatened or endangered under the Endangered Species Act', Submitted to the United States Secretary of Commerce, 21 December 2006

Cobb, J. N. (1899) 'The lobster fishery of Maine', *Bulletin of the United States Fish Commission*, vol 19, pp241–265

Cristadoro, C. (1903) 'Canvasbacks at $20 per pair', *Forest and Stream*, vol 61, p467

Edholm, C. L. (1913) 'Steaks and pearls from the abalone', *Overland Monthly*, vol 62, pp19–22

Fay, G. (1896) 'The preservation of our game and fish', *Atlantic Monthly*, vol 77, pp642–649

Inkersley, A. (1905) 'Rare shellfish', *Scientific American*, vol 93, p478

Jackson, J. B. C., Kirby, M. X., Berger, W. H., Bjorndal, K. A., Botsford, L. W., Bourque, B. J., Bradbury, R. H., Cooke, R., Erlandson, J., Estes, J. A., Hughes, T. P., Kidwell, S., Lange, C. A., Lenihan, H. S., Pandolfi, J. M., Peterson, C. H., Steneck, R. S., Tegner, M. J. and Warner, R. R. (2001) 'Historical overfishing and the recent collapse of coastal ecosystems', *Science*, vol 293, pp629–638

Lehner, J. C. (1915) *The World's Fair Menu and Recipe Book*, Lehner and Sefert, San Francisco, CA

McCusker, J. J. (2001) *How Much is that in Real Money? A Historical Commodity Price Index for Use as a Deflator of Money Values in the Economy of the United States*, American Antiquarian Society, Worcester, MA

Pauly, D., Christensen, V., Dalsgaard, J., Froese, R. and Torres, F. Jr. (1998) 'Fishing down marine food webs', *Science*, vol 279, pp860–863

Pauly, D. and Maclean, J. (2003) *In a Perfect Ocean: The State of Fisheries and Ecosystems in the North Atlantic Ocean*, Island Press, Washington, DC

Richey, E. C. (1925) 'A product of our rocky coasts', *Overland Monthly*, vol 83, p61

Rosenberg, A. A., Bolster, W. J., Alexander, K. E., Leavenworth, W. B., Cooper, A. B. and McKenzie, M. G. (2005) 'The history of ocean resources: Modeling cod biomass using historical records', *Frontiers in Ecology and the Environment*, vol 3, pp84–90

Sahr, R. C. (2005) *Inflation Conversion Factors for years 1665 to estimated 2015*, Political Science Department, Oregon State University, Oregon

Sumaila, U. R. (1999) 'Economic analyses of fisheries impacts on food webs: Pricing down marine food webs', in D. Pauly, V. Christensen and L. Coelho (eds) *Proceedings of the '98 EXPO Conference on Open Food Webs and Economic Productivity*, ACP–EU Fisheries Research Report, 5, Lisbon

USFWS (United States Fish and Wildlife Service) (2007) 'Chesapeake Field Office: Canvasback – *Aythya valisineria*', www.fws.gov/chesapeakebay/canvas.htm, accessed 12 February 2007

Ware, S. (1903) 'Curious facts and statistical truths: The abalone', *Overland Monthly*, vol 42, pp534–539

Woodard, C. (2004) *The Lobster Coast: Rebels, Rusticators and the Struggle for a Forgotten Frontier*, Viking Press, New York

Worm, B., Barbier, E. B., Beaumont, N., Duffy, J. E., Folke, C., Halpern, B. S., Jackson, J. B. C., Lotze, H. K., Micheli, F., Palumbi, S. R., Sala, E., Selkoe, K. A., Stachowicz, J. J. and Watson, R. (2006) 'Impacts of biodiversity loss on ocean ecosystem services', *Science*, vol 314, pp787–790

5

Baiting Our Memories: The Impact of Offshore Technology Change on Inshore Species Around Cape Cod, 1860–1895

Matthew G. McKenzie

In 1634, William Wood included alewives among the many gifts that nature had bestowed upon the English settlements in New England. He wrote in awe of the abundance of this anadromous fish: 'alewives ... [appeared] in such multitudes as is almost incredible, pressing up in such shallow waters as will scarce permit them to swim' (Wood, 1634, p34). Three hundred years later, Massachusetts state biologist David Belding's report on the health of the Massachusetts alewife fishery presented a more ominous picture of southern New England's former natural wealth. Even though the earliest colonial records showed that 'this fish has had considerable influence on the welfare of the country', Belding was compelled to report that: 'while in some streams the alewife fishery has held its own or even improved, it has in others to such an extent been [in] a serious decline' (Belding, 1921, pp34, 46). In just three centuries, European immigrants to New England had managed to all but destroy the once bountiful runs of alewives.

The story of how Cape Cod residents accomplished this feat challenges the current generation of marine environmental historians and historical marine ecologists to reconsider conventional explanations as to why marine resources decline. Cape Codders did not ruin their alewife runs with the onset of new market relationships. As part of the Atlantic commercial system for well over 250 years, they managed to protect these easily caught anadromous fish from the ravages of uncontrolled commercial exploitation. Nor were the runs destroyed by outsiders industrializing the fishery. Instead, changes in how local people perceived and related to their marine resources played an essential role in the destruction of the alewife fisheries. Finally, science – far from advocating a conservationist stance as it does today – played an ambiguous, catalytic role in the transition of Cape Codders from protectors to exploiters of Cape Cod alewife runs. Consequently, while Cape Cod alewife runs did collapse, this was not due to the broad changes that transformed the regional fishing industry and economy, but because of particular and localized adaptations to those changes.

Like many other marine environmental histories, this story follows a trajectory of expansion and decline: once a fishery is commercially developed, it is only a matter of time before fishers over-exploit and destroy, or all but destroy, a once thriving marine ecosystem. Key works by Arthur McEvoy and Joseph Taylor show this trend well (McEvoy, 1986; Taylor, 1999). In calling for the inclusion of humans into marine ecological understandings, McEvoy (1986) makes a solid case for also including legal and cultural understandings to explain why Californian fisheries declined so quickly. Taylor (1999) also follows this path, focusing on how the Pacific salmon fisheries contracted so quickly under commercial pressure. For both McEvoy and Taylor, the arrival of white settlers on the coast seeking easily obtained export products implicated Native Americans in new market relationships that redefined people's ties to the environment and to each other.

More recently, historical marine ecologists have adopted new approaches and methodologies to explain the history of marine resources. In 2001, Jeremy Jackson and his team of researchers published a seminal work in which historical, archaeological and palaeo-ecological sources were deployed to investigate the collapse of the Caribbean marine ecosystem (Jackson et al, 2001). In a closely interwoven mix of historical and ecological research, Rosenberg et al (2005) have also shown how hook-and-line fishers managed measurably to deplete cod stocks on the Scotian Shelf during the mid 1850s. Invoking Jackson's early ecosystem-based approach, Lotze and Milewski's (2004) innovative interdisciplinary study of the Quoddy region of the Bay of Fundy reveals a similar trend. While sustenance and sustainability could coexist before the dawn of European and Anglo-American economic relationships, an ecological downturn coincided with the early modern or modern advent of Western-style market capitalism (Lotze and Milewski, 2004).

Across this growing body of literature, commercialization emerges as a binary division between a pre-lapsarian sustainable balance between humans and nature, and the beginning of the end for marine animal populations. In other words, before the arrival of capitalist market structures – usually brought by Europeans or their Anglo-American descendants – any incentive for conservation is laid asunder by the appeal of easy profit. Both Jackson et al (2001) and Loetze and Milewski (2004) explicitly identify the arrival of Europeans as the point at which their ecosystems began to decline. For Salem/Beverly fishermen, an insatiable desire for more fish and more profits is assumed; Rosenberg et al (2005) offer no other reason to explain why fishermen changed their fishing patterns across the Scotian Shelf banks. In short, all of these studies extend the voraciousness of the modern fishing industry – indeed, of all market-based natural resource use – back into the past, never questioning whether market pressures three centuries ago differed from those prevalent today.

This is not the first time that scholars have viewed the expansion of Western European and North American capitalism and its legacies in such a binary fashion. By adopting a pre- and post-capitalist vision of decline, marine environmental historians and historical marine ecologists are presenting a picture of marine resource exploitation similar to larger images of world economic development laid out by Immanuel Wallerstein, Eric Wolf and other followers of world systems theory (Wallerstein, 1974, 1980; Wolf, 1982). Like those studying historic marine ecosystems, this interpretive sociological school sees the growth and expansion of Western-style capitalism as a pivotal event in global history. For world systems theorists, the expansion of Europe was not accomplished by competing political, religious and proto-nationalist European states, but rather represented a more unified expansion of Western European market relationships that changed people's ties to the land and to each other (DuPlessis, 1988; Shannon, 1989). In expanding, Europeans tied peripheral areas containing desirable natural resources to an economic core in an exploitative relationship. In exchange for cheap raw materials and foodstuffs produced in colonies – such as Northwest Atlantic codfish or Pacific Northwest salmon – European core areas exported expensive manufactured goods. Ultimately, this system was designed by capitalists in the core areas to maximize profits and to perpetuate an economic dependency which ensured that peripheral economies remained geared to suiting the needs of core areas. Armed with men, ships and money, Western-style market capitalism directed the world through an unstoppable, irresistible wave of economic change. This was destined to transform marine animals into commodities and to remove all barriers to over-exploitation.

The histories of many early modern fisheries tend to support the ideas and interpretations of Wallerstein and his followers. Yet, both world systems theory and the emerging marine environmental literature suffer from similar critical flaws upon which the story of Cape Cod alewives sheds some light. The most compelling criticism of world systems theory – and, by analogy, of emerging marine environmental history literature – is that by adopting a binary vision of pre-commercial and post-commercial economies, interpretation is abstracted from the particularities of each case (Chirot, 1984; DuPlessis, 1988; Shannon, 1989; Pomper, 1995; Green, 1995). In so doing, it overlooks the specific details that reveal, on a local level, how the advent of Western commercial relationships was anything but inevitable and uncontested.

Local details matter. How a community accommodates both market production and resource conservation says much about how market structures were shaped by local circumstances. And several focused studies concerned with the economic expansion of European market relationships suggest that nothing was inevitable about expanding economic systems. Richard Bushman (1998), for example, has shown that European settlers in early America viewed their agricultural and productive efforts in terms of both long-term use and market production. Bushman argues that farmers recognized that the conversion of all their resources to export goods did not necessarily fit their long-term interests, and they subsequently embraced market-oriented activities only selectively (Bushman, 1998). During the 19th century, Richard Judd (1997) has shown that in frontier Maine, it made commercial sense for settlers to preserve wilderness areas and wildlife as an insurance policy against winter hardships and crop failures. Furthermore, other scholars, such as Christopher Clark, have argued that the transformation to a more market-oriented society was anything but simple and uncontested, even among whites who had existed within the market economy for some time (Clark, 1990). Consequently, those who maintain that there was an inevitable tie between expansion of European market systems and the decline of marine resources should be careful not to overlook the particularities of time, place and history. If all politics is local, then so, too, are the politics of common resource use and abuse.

This study – or 'autopsy' (Smith and Link, 2005) – of the Cape Cod alewife fishery represents just one case in which local people kept the greatest ravages of market capitalism at bay in the face of tremendous pressures to commercialize the alewife fishery. Furthermore, once those fisheries fell under commercial use, the agent of that change took on a more familiar, and therefore more troubling, face. Sometime between 1620 and 1920 people changed how they viewed their ties to these fish and this allowed for the now familiar story of exploitation and collapse. Before the advent of highly capitalized, profit- and cod-hungry banks schooners, before vessel owners and fishermen embraced technologies that relied upon massive quantities of pre-packaged bait, and before those demands intensified inshore fishing pressure along Cape Cod shores, Cape Cod communities forgot the immediate role that these fish played in their daily lives. And with such lapses, these very real fish became abstractions – 'bait' to catch bigger game. This game included more valuable fish, new means to squeeze profit from the sea, or opportunities to prove the utility of private investment, scientific investigation and free market economics to an unmodern fishery. Consequently, far from a story of inevitable decline, Cape Codders' management and mismanagement of their local anadromous fisheries reveals the power that local people had over their fisheries resources. And this history of Cape Cod alewife runs compels us to recognize that details matter in interpreting larger trends.

As much as Cape Codders' use of alewives was driven by markets, it was also driven by alewife behaviour. For if these fish did not arrive at the time of year in which they did, and were not so easily caught, Cape Cod communities' relationship to them would stand markedly different. These 8 to 9 ounce fish, growing to 10 or 11 inches (25cm to 30cm) in length, reappear along the southern New England coast between late April and the end of June. They swim up the small brackish streams to mate and return to the sea within a few days (Collette and Klein-MacPhee, 2002, pp112–116, pp118–125). Because of the timing of the run in the early spring, when both hungry lands and hungry people needed sustenance, these fish came to play essential roles in an agricultural community's survival strategy. In addition, given the ease by which these fish could be taken, alewives were managed so that they would yield long-term benefits. The alewife fishery first fell under rigorous regulation as early as 1623 (MSC, 1887, p3). In 1651, the town of Sandwich designated two fish catchers, limited sales only to local residents, set prices and limited fishing days specifically to preserve the fishery's long-term viability (STR, 12 March, 1651/1652). Massachusetts also tried to preserve the fishery in 1709 by regulating weirs, and again in 1741 with the passage of an act explicitly designed to conserve alewives (MSC, 1887, p9). Falmouth's 1788 laws follow similar lines to those drawn by Sandwich 130 years earlier. As a common resource bestowed upon the town as a whole, only residents could benefit from these easily over-exploited fruits of nature (FTR/m, 1788). The public good of the fishery was so important that Falmouth residents went to 'war' over who should benefit from alewife streams (FTR/HF, 1790–1808). In this 1790 dispute, the town split over whether to protect the fisheries or to develop private mills. Ultimately, the issue came to an unfortunate conclusion. As town lore recounts, in the fall of 1805, Thomas Gifford pulled a relic field gun onto the town green, packed down a charge, filled it with herring and touched it off. We will never know what statement he wanted to make, as he packed down too many herring and the gun burst, killing him with the iron shrapnel. Suffice it to say, however, that Cape Cod residents took the alewife fishery seriously. During the 17th and 18th centuries, as long as Cape Codders saw alewives coming to their tables or fertilizing their fields, they had a direct tie to the natural resource and defended their rights of access and regulated pressure.

Ultimately, however, the protective attitudes of Cape Cod communities towards their respective fisheries began to wane during the late 18th century. Food supplies improved, and townspeople lost the immediate need for the early spring input of fish protein. Post-war economic recovery – taking off after 1815 – and improved coastal trading provided more reliable food deliveries to sustain the diets of Cape Codders (Kittredge, 1968, pp144–154). As Cape communities prospered, local resistance to out-of-town market sales of fish declined, and the townspeople began to look to the alewife fisheries not as food, but as revenue. As early as 1792, Rochester began selling the annual alewife fishery to the highest bidder (RTR, 5 March 1792), and six years later, Harwich voted to apply monies from the alewife fishery to pay state taxes (HTR, 8 January 1798). Sandwich, too, in 1771, looked to profits from the local alewife fishery to offset town expenses (STR, 5 March 1771). In short, affluence and comfort transformed people's relationship to their alewife runs. Rather than a tangible aid to a family's survival, benefits from the alewife fishery became income streams, breaking the bonds that had tied the welfare of the fish to the welfare of the people.

The transformation of fish into abstract revenue streams – what Cronon has called a commodity's 'second nature' – also enabled Cape Codders to respond more easily to changes in their economic world (Cronon, 1991, pp266–267).

Beginning in the early 19th century, Cape Cod offshore fleets underwent dramatic expansion and technological change. In terms of numbers, New England's offshore fleet expanded dramatically during the first half of the 19th century (McFarland, 1911, pp129–175). More specifically, changes in each fishery gave investors good reason to put their money in fishing vessels. For example, in the first third of the 19th century, the New England mackerel hook and line fishery grew dramatically as urban demand for food fish and new mackerel jigs provided both the market and the means to expand the mackerel fishery (Goode and Collins, 1887, pp187–193). In addition, Gloucester fishermen began to fish George's Bank in earnest during the 1840s to meet a new culinary demand for halibut and to tap into more valuable, previously lightly fished, stocks of cod found in those dangerous waters (Grasso, in press). Finally, across the cod fishery, fishermen began adopting tub trawls during the mid 1850s to offset catch declines (Rosenberg et al, 2005; Leavenworth, 2006).

The net result of these industrial and technological changes was the transformation of the Cape Cod alewife fisheries. Throughout the 19th century, fishermen began to understand the tastes of their prey in more sophisticated ways and, as a result, came to see fresh river herring as the choicest of baits. Armed with better bait, more vessels could fish in more areas, more efficiently landing a wider array of species (Goode and Collins, 1887, p278). With this knowledge, fishermen developed new technologies that took into account their prey's feeding behaviour. In the mackerel fishery, for example, as even small juvenile mackerel came to fetch a good market price by the 1840s, fishermen shifted to cheaper river herring species for bait. They also began 'chumming' – casting scoopfuls of ground bait into the waters around the vessel – as a means of keeping mackerel at the surface and biting hooks (Goode and Collins, 1887, p278).

In the halibut and cod fisheries, too, alewives and other river herring came to have increasingly important roles in the productive process. For both halibut and cod, tub trawls emerged as a more efficient and more bait-intensive means by which fishermen adopted a little-utilized fish, such as alewives, as a new way of increasing catches. Because hand-lining fisherman worked only two individually baited hooks at a time, bait demands were low before the 1850s. Typically, Gloucester hand-liners during the 1850s sailed with only about 500 individual herring as bait to get them started fishing. Once on the banks, bait needs could be met with gill nets run out at night (Rosenberg et al, 2005; Leavenworth, 2006). Tub-trawling, however, required more pre-packaged bait. This method used a long single line with hundreds of individual hooks attached at intervals. These hooks were baited before the set and therefore required all bait to be on board before fishermen left for the banks. By the 1880s, trawlers – now with designated bait rooms built into the vessel's very architecture – sailed with 40 to 60 barrels of bait on a Western or Grand Banks fare, while short 20-day fares to George's Bank packed 25 to 30 barrels. Even the language changed to reflect bait's importance: fares to the banks were now measured in 'baitings' – that is, the length of time that a consignment of fresh bait would last (Earll, 1887, pp439–458).

As the offshore fishery expanded, so did pressure on other species that fishermen desired for bait. Close to the fishing grounds, and with dozens of productive streams and miles of open sandy beaches, Cape Cod emerged as an important place to find bait. And this increased commercial demand for bait helped further distance Cape Codders from their alewife runs. In 1859, for example, the once protective town of Sandwich openly debated rules governing individuals' rights to sell herring (STR, 7 March 1859). By 1862, the Harwich town fishery had also fallen under an auction scheme (HTR, 4 February 1862). Taken as a whole, such schemes represented a fundamental shift in attitude about preserving fisheries. As Belding lamented in 1921:

> *The popular and easy expedient of leasing the fishery from year to year … has placed a premium upon its exploitation and has directly encouraged over-fishing. Naturally, the purchaser, uncertain of obtaining the fishery for the following years, would severely drain its resources.* (Belding, 1921, p57)

Lacking the long-term incentive for preservation, fisheries lessees took as much fish as quickly as possible to recover their lease costs and extract as much profit as they could. Conservation made little sense, for once the lease expired the fish that the lessee did not take became someone else's. Nor did town regulations and fish laws protect alewives: 'laxity in town oversight and apathetic indifference on the part of the townspeople have brought about this state of affairs' (Belding, 1921, p55). These changes in people's view of their fish – from public asset to private profit-maker – and the Cape Codders' apathy in managing their fisheries reflected how townspeople had lost their tangible connections to their alewives and, ultimately, their awareness of what was happening to them.

No longer an essential part of an annual subsistence strategy, very little prevented Cape Codders from capitalizing on their inshore fish stocks. Beginning in the 1830s, but expanding rapidly after 1850, Cape Cod residents began to sell inshore bait fish such as alewives to offshore fishing vessels needing ever increasing supplies of bait. During the first third of the 19th century, some alewife runs fell under private management for the gain of investors. In 1831, for example, Massachusetts incorporated an artificial alewife run in Harwich and Chatham to be operated as a private business. Two similar companies were incorporated in 1836, and six more before 1840. By 1863, the cape and islands were home to 23 privately managed and maintained alewife runs (MSC, 1887, pp218–273). Other fish were also being privatized in coastal waters. After the first coastal weir companies were incorporated in 1837, 1849 and 1852, 12 more were incorporated in 1855 alone (MSC, 1887, p210). Between 1865 and 1890, the number of weirs lining Cape Cod shores expanded dramatically. In 1871, Spencer Baird counted 33 weirs, traps and pounds lining the north and south shores of Cape Cod (Baird, 1873). By 1885, such installations lined the coast: in 1889 there were 97, and by 1905 128 dotted the Cape Cod coastline (Wright, 1888; USFC, 1892, p136; USDCL/BF, 1907, p49).

New fishing operations were not manifestations of outside investors exploiting a distant resource. If the Marstons Mills Fishing Co records are representative, people investing in private runs came from the same town that housed the run (CCCC, 1867–1873). Furthermore, while weir and trap operations were expensive to set up – the 82 inshore operations in 1883 had a combined value of US$89,452, or roughly US$1100 per set-up – many were owned by families such as the Atwoods, who had long been a part of the Provincetown fishing community (PMM, 1861–1877).

Landing figures detail the increased pressure on Cape Cod river herring stocks. In 1865, Cape Cod's combined take of shad, alewives, salmon and herring amounted to about 1.8 million fish. In 1885, the region took 1 million lb of alewives alone. By 1890, Cape Cod inshore operations were landing over 2 million lb of alewives (Warner, 1866; Wright, 1888; Waldin, 1899; USFC, 1892, p138). In the 1890 US Fish Commission census of the New England bait industry – the census alone testifies to the importance of industry by that time – 70 per cent of the 1.1 million lb of alewives, and over 90 per cent of the 4.6 million lb of herring landed, went for bait (USFC, 1892, p138). Between 1865 and 1895, Cape Cod's inshore fishery became a highly capitalized industry taking a lot of fish for Cape Cod investors and fishermen alike.

If this is a story of how affluence distanced a people from their natural resources, and how that perceptive distance ultimately allowed for the fisheries' over-exploitation, it is also a story about how people could justify these actions. The alewife fisheries were not simply forgotten – persistent and continued mention of these resources in town records show clearly that Cape Cod communities knew that these fisheries still existed. If people knew of the fishery, then ignorance fails to explain Cape Codders' ultimate mismanagement of their fish runs. Taylor (1999) sheds light on this in his study of Pacific salmon. Here, a misplaced faith in fish hatching allowed fishermen to take fish beyond the capacity of the ecosystem. It was not so much a question of permission, but rather an unfortunate combination of bureaucratic face-saving and a collective belief that free Americans should have the right to exploit as much as they wished (Taylor, 1999). Long before fish hatching promised falsely to solve the material problems of over-fishing, however, Cape Codders found another reason to relinquish managerial responsibility for their fisheries. Faith in free markets and private property espoused by state fisheries commissioners and local entrepreneurs alike promised cape residents and investors the best of both worlds. They could have expanded commercial benefits from their fishery and been sure that the fishery would remain viable over the long term. All cape communities had to do was abandon the traditional management practices that put the common good over private gain.

During the 1860s, Theodore Lyman, the Massachusetts Fish and Game Commissioner, had become convinced of humanity's ability to perfect nature through science, industry and free market capitalism. And as he looked at the Massachusetts inland fisheries, he identified the alewife fishery as a place where the benefits of science, law and *laissez-faire* economic theory could best be demonstrated. For him, the inshore fisheries' economic potential suffered under disincentives embedded within Cape Cod traditions of common access and benefit. Commenting on the behaviour of legislative committees considering petitions for creating private fish runs, Lyman derided the cape's commitment to traditional access rights:

> *At once the [legislators] begin to ask whether this [private] control would not abrogate some grant of the Pequot Indians to Fear-the-Lord Crowell, in the year 1639; or some ancient right of the inhabitants of Harwich Centre to dig one peck of quahogs per man on that particular ground.* (MCIF, 1868, p6)

For Lyman, such attitudes reflected the antiquated way in which governments saw property rights and fisheries resources:

> *Let our people once clearly understand that these fish ... are real property, to be increased and to be raised in value like other property, and there will be no more difficulty about the rights of owners.* (MCIF, 1868, p6)

Science – fish culture, to be precise – was the key Lyman saw to unlocking the under-optimized bounty of providence. With messianic fervour that would be familiar to Puritans such as Fear-the-Lord Crowell, Lyman saw the combination of science, private incentives and reformed laws as essential to making alewives a profitable private industry:

> *Now fish culture ... has, within these few years, become a true INDUSTRY, one that should bring wealth to individuals and revenue to the Commonwealth. It asks only ... that our inland waters should, in large spirit, be thrown open to the energy of private persons; for with us it must always be private persons, and not the government, that build up new industries.* (MCIF, 1868, p22)

For Lyman and investors in the growing coastal fishing operations of the 1860s, the annual alewife runs represented a means by which private interests could demonstrate the benefits of *laissez-faire* economics and science over antiquated traditions of common access and common benefit. Like Taylor's story of the Pacific salmon fisheries, science catalysed the over-exploitation of the Cape Cod alewife runs, but did so in an indirect way. Cape Cod witnessed no influx of foreign capital, despite hook-and-line fishermen's claims to the contrary, nor did the Cape witness outsiders entering communities and buying up fishing operations. Unlike the Pacific Northwest, where fishing companies sent operatives to take fish, Cape Codders industrialized their own fisheries for their own profit. In this case, science allowed Cape Codders to justify abandoning traditions that limited fishing pressure, guaranteed local access and ensured that all in the community – but only the community – benefited from local natural bounty.

Far from a story of distant market systems invading an area and imposing new relationships upon people and the land, the story of Cape Cod alewives suggests that such a process followed more complex and contingent routes. For two centuries, Cape Codders were tied to their spring runs of alewives. Whether for food or fertilizer, alewives played an important role in Cape Codders' annual subsistence strategies. This changed as communities became more nutritionally stable through expanded trade networks, and as expanding offshore fleets embraced new bait-intensive technologies. At that point, with their fish merely abstractions on town account books, Cape Cod communities relinquished their responsibility to preserve the long-term health of their fisheries. Justified by science and the promise of *laissez-faire* economics, Cape Codders destroyed the fisheries by allowing themselves to see alewives only as inputs into larger productive processes.

Ironically, little changed when Belding issued his 1921 report. Far from a halcyon cry to conservation for its own sake, Belding called for alewife restoration based upon the same instrumental grounds that led to the fishery's destruction in the first place. He reported: 'few more convincing arguments can be advanced in favour of replenishing all possible alewife streams than its influence upon the future welfare of the shore fisheries of Massachusetts' (Belding, 1921, p46). Rather than representing a resource worthy of conservation on its own terms, the precarious state of the alewife fishery in 1921 threatened the viability of other fisheries, and the importance of these fish – as bait – was why Massachusetts should restore the stocks. Nor did Cape Codders awaken to the plight of their fish. In perhaps the first time in the history of New England fisheries management, no one – not a single organization – complained when Massachusetts, Connecticut and Rhode Island closed their river herring fisheries in 2006. According to one fisheries manager, this silence represented indifference rather than a universal acceptance of the need to close the fishery. Put simply, no one cared if the fishery was open or closed.

On the surface, this story, like many other marine environmental histories, fits into a larger picture of expanding market systems forever changing the relationships between people and natural resources. Such an image is well supported by solid, empirically based historical and scientific studies that, when assembled as a whole, will present a picture of marine animal populations in the past. In addition, these studies fit into other understandings of world history that further confirm the importance and validity of the History of Marine Animal Populations project's (HMAP's) unique approach. This little story of a little fish from a little area demonstrates, however, that surface impressions bury a host of details that reveal important contingencies essential in understanding how marine animal populations evolved as they did. It would be easy to see this as another tale of collapse following the development of capitalist market relations, technology change and the industrialization of the fishery. But we cannot use a simple binary vision of pre-capitalist and post-capitalist change to explain the demise of Cape Cod's alewives. This history shows that local peculiarities – in this case, affluence, apathy and a new faith in science and industry – played as important a role in precipitating decline as the power of outside forces. Consequently, we need to rethink how we piece these local stories together into a meaningful pattern, but one that does not 'lose the fish for the school'.

REFERENCES

Baird, S. F. (1873) *Report on the Condition of the Sea Fisheries of the South Coast of New England in 1871 and 1872*, Government Printing Office, Washington, DC

Belding, D. (1921) A *Report upon the Alewife Fisheries of Massachusetts*, Wright and Potter, Boston, MA

Bushman, R. L. (1998) 'Markets and composite farms in early America', *William and Mary Quarterly*, (3rd series), vol 55, pp351–374

CCCC (Cape Cod Community College) (1867–1873) *Marstons Mills Fishing Company Records*, Nickerson Room, Barnstable, MA

Chirot, D. (1984) 'Review of Eric Wolf, Europe and the people without history', *Journal of Social History*, vol 18, pp119–124

Clark, C. (1990) *The Roots of Rural Capitalism: Western Massachusetts, 1780–1860*, Cornell University Press, Ithaca, NY

Collette, B. B. and Klein-MacPhee, G. (eds) (2002) *Bigelow and Schroeder's Fishes of the Gulf of Maine*, Smithsonian Institute Press, Washington, DC

Cronon, W. (1991) *Nature's Metropolis: Chicago and the Great West*, W. W. Norton & Co, New York, NY

DuPlessis, R. S. (1988) 'Wallerstein, world systems analysis, and early modern European history', *The History Teacher*, vol 21, pp221–232

Earll, E. (1887) 'The herring fishery and the sardine industry', in G. B. Goode (ed) *The Fisheries and Fishery Industry of the United States. Section V, vol I: History and Methods of the Fisheries*, Government Printing Office, Washington, DC

FTR/HF (1790–1808) *Herring Fishery, 1790–1808*, Falmouth Town Records, Falmouth Historical Society, Falmouth, MA

FTR/m (1788) Falmouth Town Records on microfilm, Falmouth Public Library, Falmouth, MA, 18 March

Goode, G. B. and Collins, J. W. (1887) 'The mackerel fishery', in G. B. Goode (ed) *The Fisheries and Fishery Industry of the United States. Section V, vol I: History and Methods of the Fisheries*, Government Printing Office, Washington, DC

Grasso, G. M. (in press) 'What appeared limitless plenty: The rise and fall of the nineteenth century Atlantic halibut fishery', *Environmental History*

Green, W. A. (1995) 'Periodizing world history', *History and Theory*, vol 34, pp99–111

HTR (dates as given) Harwich Town Records, Harwich Town Clerk's Office, Harwich, MA

Jackson, J. B. C., Kirby, M. X., Berger, W. H., Bjorndal, K. A., Botsford, L. W., Bourque, B. J., Bradbury, R. H., Cooke, R., Erlandson, J., Estes, J. A., Hughes, T. P., Kidwell, S., Lange, C. A., Lenihan, H. S., Pandolfi, J. M., Peterson, C. H., Steneck, R. S., Tegner, M. J. and Warner, R. R. (2001) 'Historical overfishing and the recent collapse of coastal ecosystems', *Science*, vol 293, pp629–638

Judd, R. (1997) *Common Lands, Common People: The Origins of Conservation in Northern New England*, Harvard University Press, Cambridge, MA

Kittredge, H. C. (1968) *Cape Cod: Its People and Their History*, Houghton Mifflin & Co, Boston, MA

Leavenworth, W. B. (2006) 'Opening Pandora's Box: Tradition, competition and technology on the Scotian Shelf, 1852–1860', in D. J. Starkey and J. E. Candow (eds) *The North Atlantic Fisheries: Supply, Marketing and Consumption, 1560–1990*, Maritime Historical Studies Centre, University of Hull, UK

Lotze, H. and Milewski, I. (2004) 'Two centuries of multiple human impacts and successive changes in a North Atlantic food web', *Ecological Applications*, vol 5, pp1428–1447

McEvoy, A. M. (1986) *The Fisherman's Problem: Ecology and Law in the California Fisheries, 1850–1980*, Cambridge University Press, Cambridge, UK

McFarland, R. (1911) *A History of the New England Fisheries*, University of Pennsylvania, New York, reprinted in 2002 by Martino Publishing, CT

MCIF (Massachusetts Commissioners of Inland Fisheries) *Reports of the Commissioners of Inland Fisheries, 1857–1871*, Wright and Potter, Boston, MA

MSC (Massachusetts, Secretary of the Commonwealth) (1887) *Laws Relating to the Inland Fisheries in Massachusetts, 1623–1886*, Wright and Potter, Boston, MA

PMM (Pilgrim Monument Museum) (1861–1877) *Inshore Fishery Day Books, 1861–1877*, Atwood Co Collection, Provincetown, MA

Pomper, P. (1995) 'World history and its critics', *History and Theory*, vol 34, pp1–7

Procter Brothers (1882) *The Fishermen's Own Book*, Proctor Brothers, Gloucester, MA

Rosenberg, A. A., Bolster, W. J., Alexander, K. E., Leavenworth, W. B., Cooper, A. B. and McKenzie, M. G. (2005) 'The history of ocean resources: Modelling cod biomass using historical records', in *Frontiers in Ecology and the Environment*, vol 3, pp84–90

RTR (date as given) Rochester Town Records, Rochester Town Hall, Rochester MA

Shannon, T. R. (1989) *An Introduction to the World-System Perspective*, Westview Press, Boulder, CO

Smith, T. and Link, J. S. (2005) 'Autopsy your dead ... and living: A proposal for fisheries science, fisheries management and fisheries', *Fish and Fisheries*, vol 6, pp73–87

Steinberg, T. (1991) *Nature Incorporated: Industrialization and the Waters of New England*, University of Massachusetts Press, Amherst, MA

STR (dates as given) Sandwich Town Records, Sandwich Public Library, Sandwich, MA

Taylor, J. E. (1999) *Making Salmon: An Environmental History of the Northwest Fisheries Crisis*, University of Washington Press, Seattle, WA

USDCL/BF (United States Department of Commerce and Labor, Bureau of Fisheries) (1907) *Statistics of the Fisheries of the New England States for 1905*, Government Printing Office, Washington, DC

USFC (United States Fish Commission) (1892) 'Report on the fisheries of the New England states', *Bulletin of the United States Fish Commission for 1890*, vol 10, pp73–176

Waldin, H. G. (1899) *Census of the Commonwealth of Massachusetts, 1895, vol VI: The Fisheries, Commerce and Agriculture*, Wright and Potter, Boston, MA

Wallerstein, I. (1974) *The Modern World-System I: Capitalist Agriculture and the Origins of the European World-Economy in the Sixteenth Century*, Academic Press, New York, NY

Wallerstein, I. (1980) *The Modern World-System II: Mercantilism and the Consolidation of the European World-Economy, 1600–1750*, Academic Press, New York, NY

Warner, O. (1866) *Statistical Information Relating to Certain Branches of Industry in Massachusetts for the Year Ending May 1, 1865*, Wright and Potter, Boston, MA

Wolf, E. (1982) *Europe and the People without History*, University of California Press, Berkeley, CA

Wood, W. (1634) *New England's Prospect*, John Bellamie, London, UK

Wright, C. D. (1877) *The Census of Massachusetts: 1875, vol II: Manufactures and Occupations*, Albert J. Wright, Boston, MA

Wright, C. D. (1888) *Census of Massachusetts, 1885, vol II: Manufactures, The Fisheries, and Commerce*, Wright and Potter, Boston, MA

6

Mapping Historic Fishing Grounds in the Gulf of Maine and Northwest Atlantic Ocean

Stefan Claesson

The Gulf of Maine Cod Project (GMCP), based at the University of New Hampshire (UNH), is a research initiative conducted under the auspices of the History of Marine Animal Populations (HMAP) programme. Its general aims are to deploy historical evidence to better understand the causes and consequences of long-term changes in marine environments, and to establish biological indicators and population trends for fish stocks in the Northwest Atlantic. More specifically, GMCP addresses three research questions. First, to what extent can historical data identify and define marine habitats? Second, what were the patterns that characterized, and the processes that shaped, the diversity, population and distribution of fish species in the past? Third, how have fishing communities perceived the nature and causes of change in marine ecosystems and how have they adapted to new environmental and regulatory conditions? In adopting a case study approach to answering these questions, GMCP's initial research has focused upon analysing fishing records relating to two specific regions: the Scotian Shelf and the Gulf of Maine (see Figure 6.1).

Hundreds of fishermen's logbooks and fishing agreements dating from the 1850s and 1860s exist in libraries and archives throughout New England. From these, GMCP has extracted qualitative and quantitative data in order to estimate the abundance of marine life in the regions under examination. In particular, GMCP has used these records to estimate the biomass of cod on the Scotian Shelf. This region was one of the most prolific fisheries in the world; but during the 1990s, it witnessed one of the worst declines in groundfish populations ever recorded, especially with regard to cod and halibut (Haedrich and Hamilton, 2000; Myers and Worm, 2003). Using information gleaned from the logbooks, GMCP

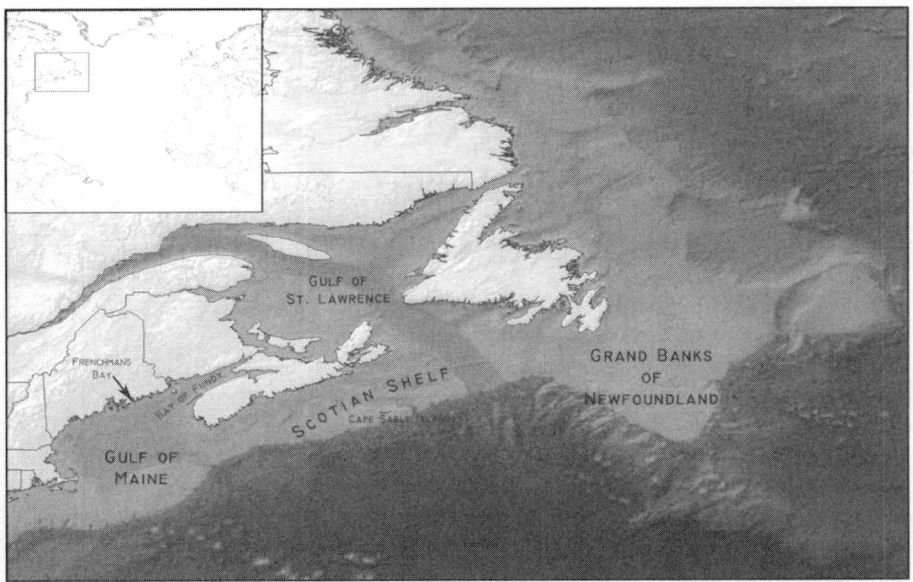

Figure 6.1 *The study area of the Gulf of Maine Cod Project*

Note: The study area of the Gulf of Maine Cod Project comprises the Gulf of Maine, Bay of Fundy, Scotian Shelf, Gulf of St Lawrence and the Grand Banks of Newfoundland.

Source: Bathymetry and topography are derived from '2-Minute Gridded Global Relief Data' (ETOPO2) (www.ngdc.noaa.gov/); view colour version at www.hull.ac.uk/oceanspast

has estimated that adult cod biomass on the Scotian Shelf was at least 1.26 million metric tonnes in 1852 (Rosenberg et al, 2005). The biomass of this region today is approximately 50,000 metric tonnes. This marked contraction, whether caused by over-fishing, climate change or pollution, has upset the diversity and health of one of the world's most productive ecosystems. This, in turn, has impacted negatively upon the economic, social and cultural stability of coastal communities in the northeast US and the Canadian Maritimes (Lotze et al, 2006, p1806). GMCP endeavours to unravel how, why and where such declines in stocks occur, and to analyse the societal and cultural consequences of marine environmental degradation.

HISTORICAL SOURCES

The Gulf of Maine and the Northwest Atlantic were explored and fished by Europeans well before any settlement was firmly established along the North

American coast during the early 17th century. Historical records indicate that the Northwest Atlantic has been fished commercially since at least 1502, and archaeological evidence confirms that Basques were whaling and cod-fishing in northeastern Canada by the mid 1500s (Pope, 2004, Innis, 1940; Grenier, 1988). The population of marine animals prior to European settlement and resource exploitation is difficult to determine, if not impossible to imagine. This is particularly true in light of early accounts that point to the existence of a somewhat unbelievable abundance of fish in the Gulf of Maine. For instance, in 1602, on returning from a voyage to Cape Cod, Bartholomew Gosnold reported:

> *In five or six hours absence, we had pestered our ship so with Cod fish, that we threw numbers of them over-board again: and surely, I am persuaded that in the months of March, April, and May, there is upon this coast, better fishing, and in as great plenty, as in Newfoundland: for the shoals of mackerel, herrings, Cod, and other fish, that we daily saw as we went and came from the shore, were wonderful; and besides, the places where we took these Cods (and might in a few days have laden our ship) were but in seven fathom water, and within less than a league of the shore; where, in New-found-land they fish in forty or fifty fathom water, and far off.* (Brereton, 1903, pp4–5)

Three years later, James Rosier was equally effusive about the region's fish stocks:

> *We were so delighted to see them catch so great fish, so fast as the hooke came down, some with playing with the hooke they tooke by the backe, and one of the mates with two hookes at a lead at five draughts together haled up tenne fishes. All were generally very great, some they measured to be five foot long and three foot about.* (Burrage, 1906, p390)

In a similar vein, Emmanuel Altham, a settler from Plymouth, Massachusetts, commented:

> *And now to speak more at length of the country and what profit is to be raised here; and first, to speak of the fishing that is in the country. Indeed it is beyond belief but I can assure you this much: that if a man be well provided with all the things necessary, for to make a fishing voyage, he may easily make his voyage two for one. But lest this should seem incredible to some, I will give you an instance of the same. Before we got hither to Patuxet, we had many fogs at sea, insomuch that we were driven to lie at hull and to take down all our sails, and so went to fishing. In one hour we got 100 great cod, and if we would have but stayed after the fog broke up, we might quickly loaded our ship and, I am persuaded, had we been prepared with all things necessary for a*

> *fishing voyage, we might have loaded our ship in a week. I think we got 1000 in all. And indeed, when we had nothing else to do, my people took delight to catch them, although we threw them away again, as I think we did 300. One fish we got, I think, weighed 100 pound: it was as big a cod as ever was seen. We got many turbots, likewise, and one turbot we got gave all our ship a meal and to spare.* (Altham, 1623)

Historians have generally dismissed early descriptions of marine abundance as embellishments used by early explorers to promote settlement or secure funding for future expeditions to the New World (Palfrey and Palfrey, 1865, pp89–98). Such accounts of fish abundance are also generally viewed as unreliable from a scientific perspective and discounted in current ecosystem evaluations. However, such qualitative sources, particularly when corroborated by archaeological, palaeo-ecological and other forms of physical data, and subjected to modelling techniques (e.g. statistical and spatial) can be used to establish estimates of abundance and distribution in past fish populations. Moreover, these narratives can also contribute to a clearer understanding of the long-term anthropogenic impacts upon marine ecosystems (Pauly, 1995). Applying geographic information systems (GIS) methods to qualitative, as well as quantitative, information, as this chapter demonstrates, facilitates the mapping of historic fisheries data. It also suggests that early, eye-witness accounts of fantastic fish abundances are often not far removed from reality.

Two varieties of primary source material were sampled for this fisheries mapping investigation. One comprised fishing charts published by the US Commission on Fish and Fisheries (Goode, 1887; Rich, 1929). The other consisted of logbooks that were generated by the cod-fishing fleets despatched from the Massachusetts ports of Beverly and Salem during the mid 19th century. The fishing charts are well known among historians and marine biologists, although prior to GCMP they had not been available in a digital format suitable for GIS analysis. Of the 2000 or so New England logbooks that have survived from the 1850s and 1860s, a sample of 27 was used in this study to test the potential of this extraordinarily rich primary source for the assessment of fish population trends over time.

Fishing charts

The fishing ground charts digitized for this study are found in two US government publications: *The Fisheries and Fishery Industries of the United States* (Goode, 1887) and the *Fishing Grounds of the Gulf of Maine* (Rich, 1929). Goode's survey of the US fishing industry, which was sponsored by the US Fish Commission and the Census Bureau during the 1880s, resulted in the publication of an epic work comprising seven volumes. It includes a comprehensive study of all known fishes, marine mammals and invertebrates exploited by the 19th-century US fishing industry. It also illustrates the locations of fishing grounds, documents fishing methods, describes the labour force in terms of the cultural affiliations and ethnicity of fishers, and provides detailed statistics of the industry in the late 19th century. The report submitted to the Bureau of Fisheries by Walter Rich, though not as comprehensive as the Goode report, provides detailed charts of fishing grounds in the Gulf of Maine and the Bay of Fundy, together with a comprehensive list of species fished on each ground.

Section III ('The fishing grounds of North America with forty-nine charts') of the Goode (1887) report includes 12 charts, drawn at varying scales, of the Northwest Atlantic and Gulf of Maine (see Figure 6.2). Features illustrated on

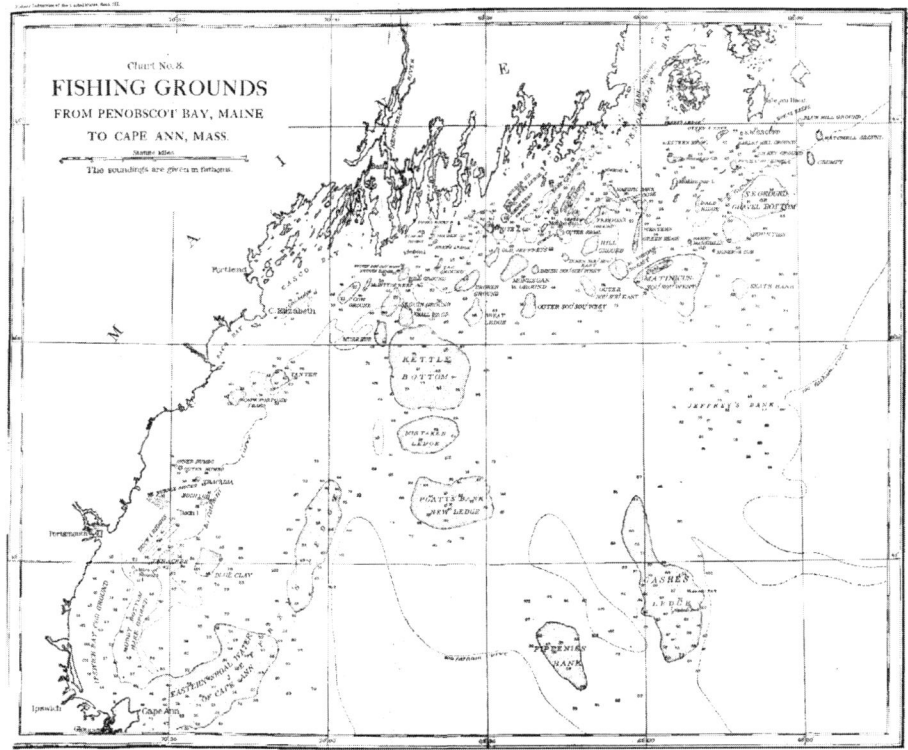

Figure 6.2 *Chart No 8: Fishing grounds from Penobscot, Maine, to Cape Ann, Massachusetts*

Note: This is one of 12 Northwest Atlantic Ocean charts published in *The Fisheries and Fishery Industries of the United States* (Goode, 1887). These charts illustrate all fishing grounds in the region known to American fishermen by the mid to late 19th century.

Source: Goode (1887, pp30–31)

the charts include fishing grounds and banks, points of interest – such as large catch reports, sightings of giant squid and locations of deep-sea coral – and areas where fish had become locally extinct. Goode relied upon published charts, as well as information provided by fishermen, to define fishing banks and grounds. The accuracy and determination of the boundaries of the fishing grounds and banks vary slightly for inshore and offshore fisheries. With regard to the former areas:

> ... it is not expected that absolute accuracy has been reached in all cases, especially as regards distances, the fishermen having to rely mainly on their judgement in this matter, it is probable that the positions and outlines assigned to the grounds are, in the main, correct. (Goode, 1887, px)

In the case of grounds located offshore:

> ... fishermen make use of a patent log for determining distances and also ascertain their positions by observations for latitude and longitude, and their statements respecting localities at a distance from shore are as reliable as could be obtained from any source. (Goode, 1887, pxi).

The report also states that the contours of offshore banks were based upon British Admiralty charts for the areas from Brown's Bank to the Flemish Cap. Contours for George's Bank were derived from US Coast Survey maps, as well as data provided by US Fish Commission officers.

There are a total of five charts published in Appendix III of the Rich report (see Figure 6.3). The charts are drawn at varying scales and detail. To illustrate the banks and ledges of the Gulf of Maine and Bay of Fundy, Rich referred mainly to published charts (such as the US Coast Survey and British Admiralty charts). Most of the grounds, however, were mapped by consulting 'large numbers of fishing captains' (Rich, 1929, p53). Although numerous grounds identified by Rich are not illustrated on the Goode charts, Rich used the Goode reports as a baseline from which he documented any significant changes in environmental conditions, fishing methods and the type of species caught on each ground.

Fishing logbooks

In 2001, UNH graduate students and researchers uncovered a wealth of Customs House records – including fishing logbooks, agreements, licences, bonds and enrolments – at the National Archives and Records Administration (Waltham, Massachusetts) and James Duncan Phillips Library at the Peabody-Essex Museum (Salem, Massachusett). While logbooks exist for numerous fishing ports throughout New England, the Beverly logbooks appear to represent the most complete

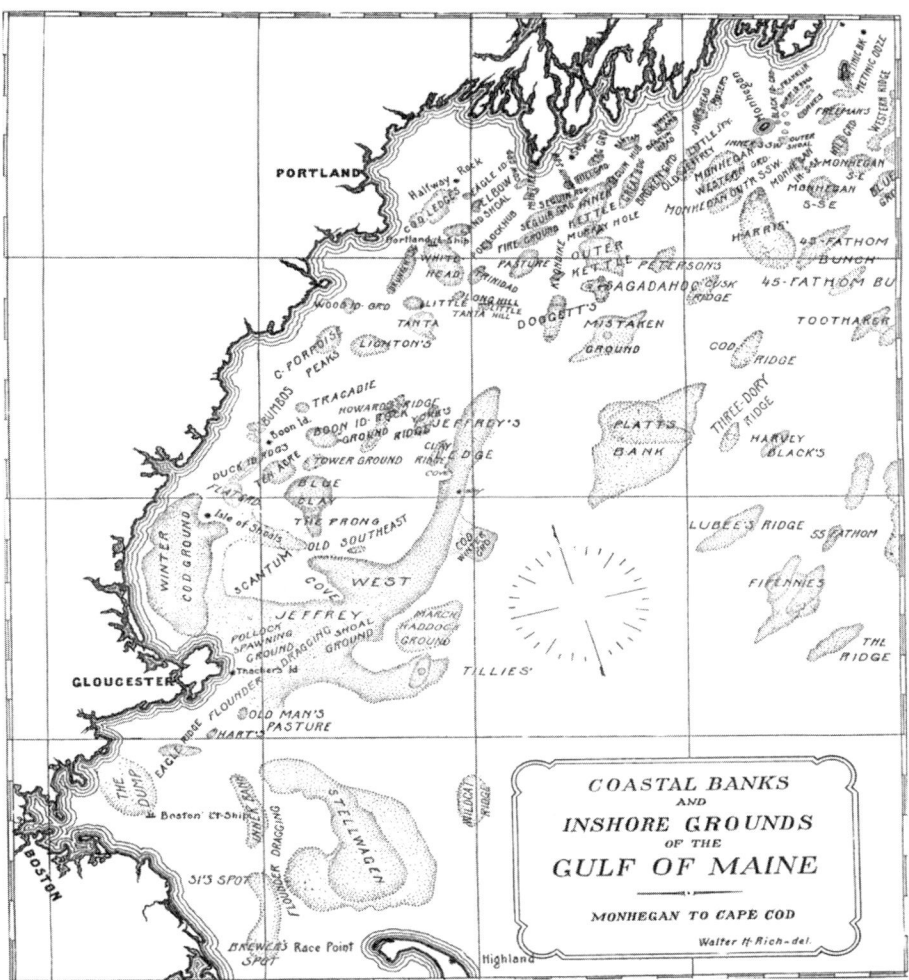

Figure 6.3 *Coastal banks and inshore grounds of the Gulf of Maine, Monhegan to Cape Cod*

Note: This is one of five detailed charts of the Gulf of Maine and Bay of Fundy fishing grounds printed by the US Bureau of Fisheries and Walter Rich in 1929.

Source: Rich (1929, pp74–75)

collection. A total of 27 of the 710 logbooks from Beverly were transcribed for this study. Spanning eight years (1852 to 1860), these logbooks were created during a period when Beverly's fishing effort was focused exclusively on the cod stocks of the Scotian Shelf, which consisted of grounds known as Banquereau, Sable Island Bank (i.e. Western Banks) and Middle Ground.

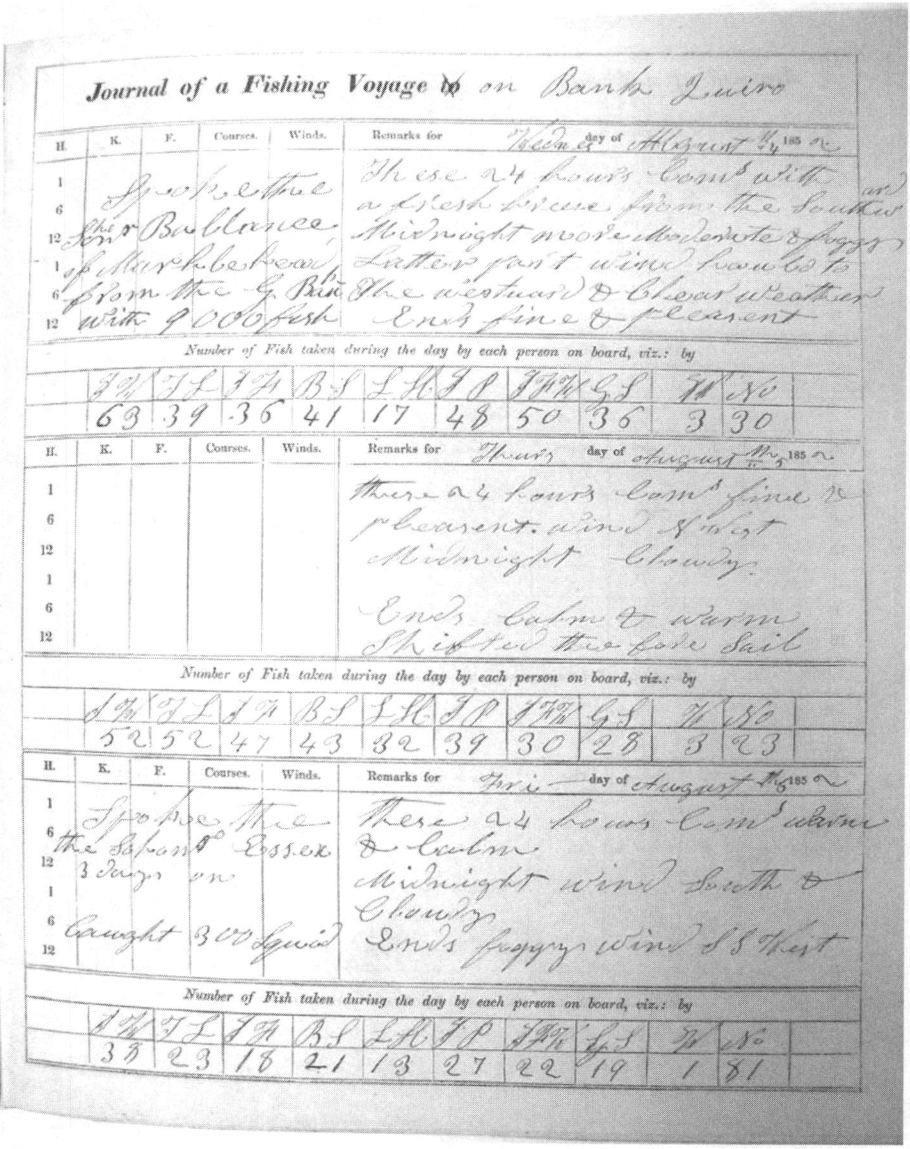
Journal of a Fishing Voyage to on Bank Quiro

H.	K.	F.	Courses.	Winds.	Remarks for Wednes day of August 4 185 2
1 6 12 1 6 12	Spoke the Schr Ballance of Marblehead from the G. Bank with 9000 fish				These 24 hours Comes with a fresh breeze from the Southward Midnight more moderate & foggy Latter part wind hauled to the westward & Clear weather Ends fine & pleasent

Number of Fish taken during the day by each person on board, viz.: by

J W	T L	J H	B S	L H	J P	J Wh	G S	W	No
63	39	36	41	17	48	50	36	3	30

H.	K.	F.	Courses.	Winds.	Remarks for Thurs day of August 5 185 2
1 6 12 1 6 12					These 24 hours Comes fine & pleasent. wind N West Midnight Cloudy. Ends Calm & warm Shifted the fore Sail

Number of Fish taken during the day by each person on board, viz.: by

J W	T L	J H	B S	L H	J P	J Wh	G S	W	No
52	52	47	43	32	39	30	28	3	23

H.	K.	F.	Courses.	Winds.	Remarks for Fri day of August 6 185 2
1 6 12 1 6 12	Spoke the the Schoonr Essex 3 days on Caught 300 Squid				These 24 hours Comes warm & Calm Midnight wind South & Cloudy Ends foggy wind S S West

Number of Fish taken during the day by each person on board, viz.: by

J W	T L	J H	B S	L H	J P	J Wh	G S	W	No
33	23	18	21	13	27	22	19	1	81

Figure 6.4 *A page from the logbook of the Beverly, Massachusetts, fishing schooner* Dove *(1854)*

Note: Logbooks typically provide a summary of weather conditions, vessels 'spoken to' at sea, vessel position (in latitude and longitude) or fishing location (e.g. 'Bank Quiro', or Banquereau), and the number of fish caught by individual fishermen each day.

Source: National Archives and Records Administration, Waltham, MA

The logbooks provide a detailed daily record of observations at sea, including vessel name, date, weather and wind conditions, position (latitude and longitude), water depth, crew composition, type and number of marine species caught, fishing technology, unusual events and notes on communications with other vessels (see Figure 6.4). They were generated according to federal bounty regulations that required vessels engaged in the cod fishery to document their time and activity at sea. Initially, the US Congress passed a law in 1792 that offered a bounty or subsidy to the cod-fishing industry in order to offset the high tariffs placed on imported salt. Vessels had to fish exclusively for cod for at least four months during the year in order to receive the bounty. An annual allowance was paid according to vessel tonnage, with its owner and crew receiving three-eighths and five-eighths, respectively. The fishermen divided their allowance in line with their share of the annual catch (O'Leary, 1996, p41). Although logbooks were occasionally falsified so that the bounty would be paid without fulfilling the requisite time cod fishing, there was little incentive to misrepresent daily catch records (O'Leary, 1996, p60). In 1852, numerous rules and regulations were amended to the bounty law by the Secretary of the Treasury. One of these rules included keeping a formal logbook while at sea. The cod-fishing bounty by the 1860s, while an economic necessity for the survival of small fishing enterprises and working fishermen, was viewed as a regulatory burden by the larger fishing capitalists. Consequently, it was repealed in 1866, its legacy being an exceptionally detailed source of fisheries data spanning the 1852 to 1866 period.

METHODS

A total of 17 historical fishing charts from the Goode and Rich publications were scanned and geo-referenced. The 27 Beverly logbooks – referred to here as Daily Catch Records (DCRs) – were also transcribed and entered into a tabular database. Vector- and raster-based data products were developed from these sources. Each dataset required a slightly different processing technique, as described below. Metadata documentation was developed for each GIS dataset according to the Federal Geographic Data Committee's (FGDC) Content Standards for Digital Geospatial Metadata (CSDGM, FGDC-STD-001-1998). These afford users a description of the contents and the original source material, together with an explanation of the digitization and database development process. The metadata also provide the historical context needed for accurate interpretation of the data.

Recreating the geography of the 19th-century grounds

The Rich and Goode fishing charts were scanned as greyscale images at 300 dots per inch (dpi). They were subsequently cleaned of background noise and cropped to the borders of the charts. Image editing software Adobe Photoshop 7.0 was then used to adjust the contrast and light balance. The images were saved as uncompressed .tiffs. In order to orient the chart to real-world coordinates using geo-referencing tools in ESRI's ArcGIS 9.0, a minimum of four reference points on each image were selected. The images were geo-referenced through graticules (in latitude and longitude) drawn on the original charts. The four corners of each chart were typically chosen as reference points. Whenever possible, however, graticule intersections at the centre of the charts were also chosen as reference points. The images were geo-referenced using first- or second-order polynomial transformations. The geographic projection of the raster charts was set to the 1984 World Geodetic System (WGS84), and the images were exported and saved as GeoTIFF.

The charts were then imported into the ArcGIS 9.0 user interface ArcMap following scanning and geo-referencing. Freely available Vector Map Level 0 (VMAP) political boundary data (1:1,000,000) of the Northwest Atlantic Region (northeast US and Canadian provinces) were used both as base layers and geographic orientation. The fishing grounds and banks illustrated on the raster charts were then digitized ('heads up', or on screen) as polygons. Because the charts were drawn at varying scales, there are usually horizontal inaccuracies where the charts overlap. Consequently, preference was always given to the chart with the larger scale and greater detail when digitizing. Polygons that abutted or intersected the shore were 'clipped' to the VMAP coastline data. A text field was added to each vector database (e.g. GROUND_ID), which identifies the name of each fishing ground, bank and ledge on the charts.

The geo-referenced fishing charts recreate the geography known by 19th-century American fishermen, revealing spatial characteristics noted systematically by fishermen before the era of scientific fish sampling (see Figure 6.5). As well as illustrating the fishing banks and grounds in the Northwest Atlantic, the Goode charts identify undersea features (e.g. deep-sea coral), points where large catches of fish were reported, seasonal fishing areas and locations where certain species went locally extinct. These features were digitized in separate vector data layers as lines, points and polygons. Attribute data fields were added to code the type of point data represented (i.e. 1 = large catch; 2 = deep-sea coral; 3 = notation). The 'notation' code refers specifically to Roman numerals illustrated on the charts for which site descriptions are provided in the Goode report. These notations refer to unusual events or sightings at sea, such as the presence of giant squid, as well as reports of unusually large and quantifiable fish catches. Line data digitized from the charts typically follow a length of coastline or fishing bank that report where the best catches were found, as well as the localities where fish species went extinct (see Figure 6.6). A polygon layer was also digitized from the charts to indicate areas identified as the best fishing areas, grounds fished for certain species and seasonal fishing locations. An attribute field containing text descriptions of chart notations was appended to each database. Finally, each vector dataset was exported as an

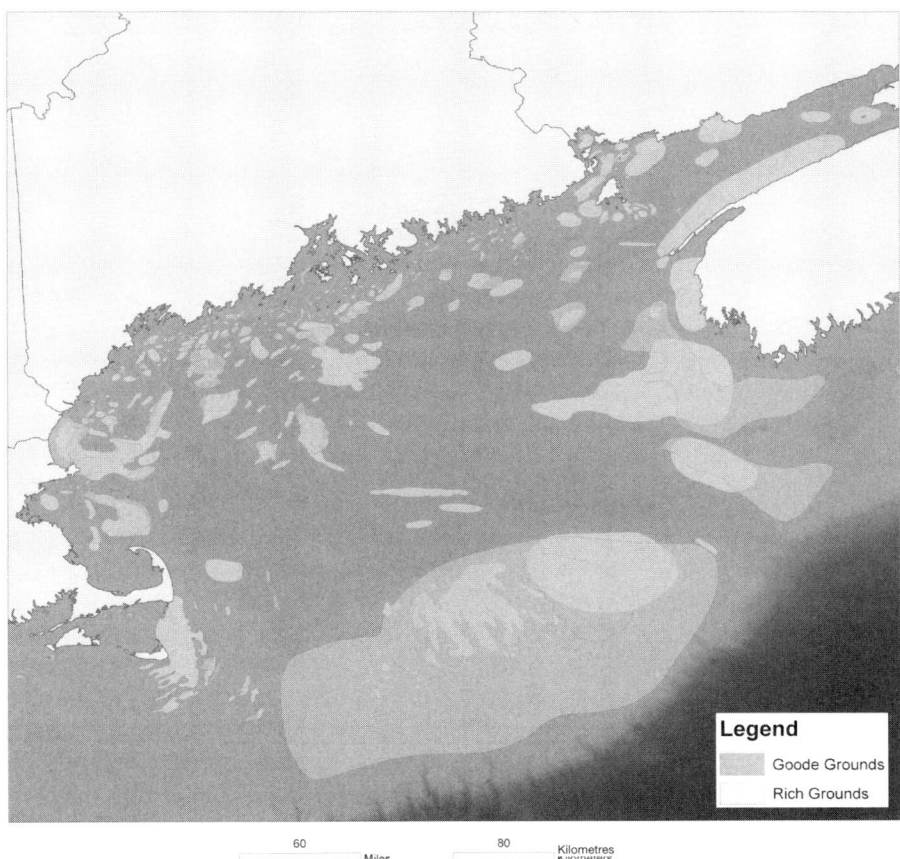

Figure 6.5 *Fishing charts of known and exploited fishing grounds and banks in the Gulf of Maine during the late 19th and early 20th centuries*

Source: derived from Goode (1887); Rich (1929); view colour version at www.hull.ac.uk/oceanspast

ESRI Shapefile, and metadata records were created in ArcCatalog, a user interface of ArcGIS 9.0.

Daily Catch Records (DCRs)

DCRs contain extensive information about day-to-day fishing effort in the Gulf of Maine and Northwest Atlantic. Data fields in the DCR dataset include the logbook entry date; the vessel position in latitude and longitude; the name of the ground or bank fished; the number of fish caught at location; the size of fish caught, usually in terms of 'small', 'large' and 'very large'; the technology or gear employed for fishing; and a description of bottom conditions. The material also

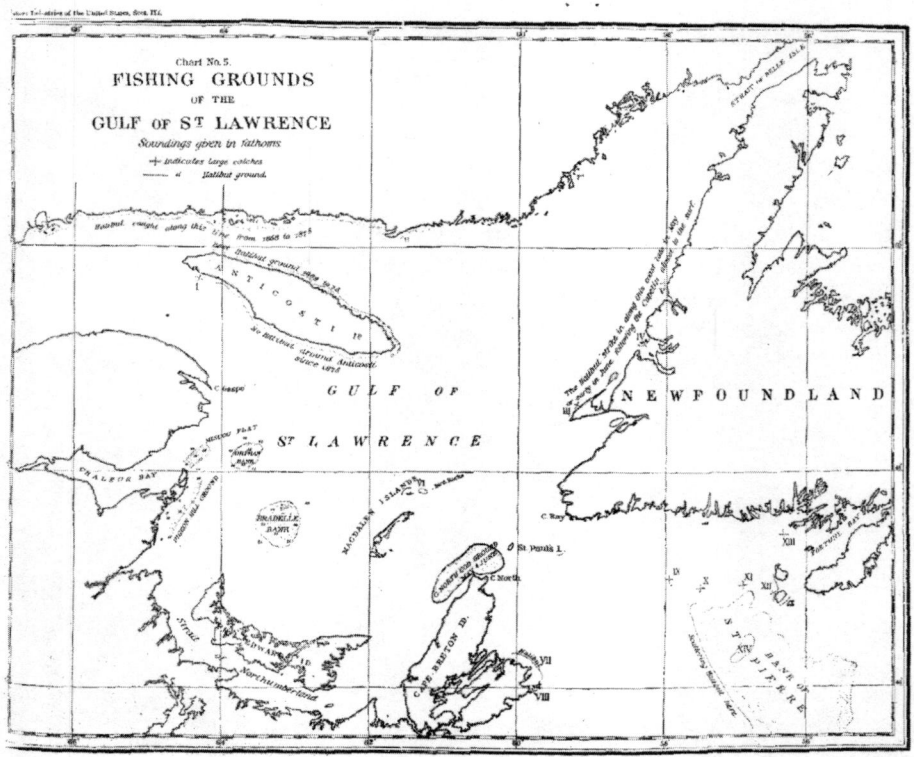

Figure 6.6 *Goode Fishing Chart No 5: Fishing Grounds of the Gulf of St Lawrence (1887)*

Note: This chart records those inshore fisheries that were depleted by the 1870s; Anticosti Island was the 'best halibut ground 1869 to 1875', but there was 'no halibut around Anticosti since 1875'.

Source: Goode (1887, pp16–17); see text; view colour version at www.hull.ac.uk/oceanspast

includes the logbook compiler's general observations at sea and in port, such as unusual natural phenomena, vessels spoken to at sea and vessel outfitting. DCRs were transcribed into Microsoft Excel spreadsheets, and prior to importing the DCR data into ArcGIS 9.0, attribute fields containing latitude and longitude coordinates were converted from their original degrees and minutes format to decimal degrees. The vessel positions or data points were then visually checked for possible transcription or data entry errors. The projection and coordinate system of the data were defined (WGS84) and the resulting geospatial data were exported as ESRI-formatted Shapefiles.

The small sample of Beverly logbook data mapped in this study suggests that previously unavailable local population estimates could be derived from

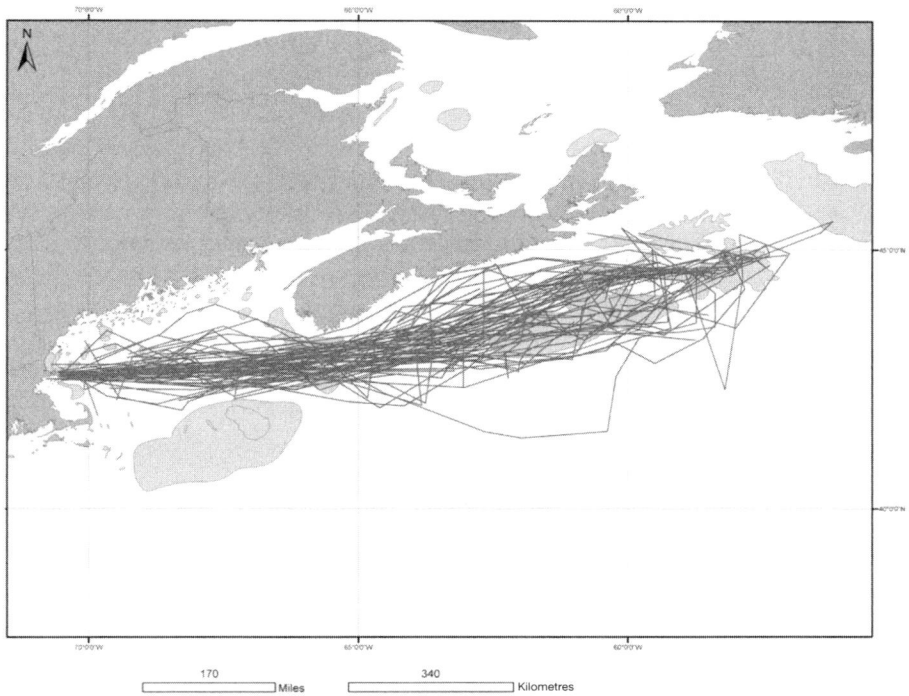

Figure 6.7 *Sailing courses of 27 American schooners from Beverly and Salem, Massachusetts, that fished on the Scotian Shelf during the 1850s*

Source: Fishing Logbooks, National Archives and Records Administration, Waltham, MA; view colour version at www.hull.ac.uk/oceanspast

these historical sources. Data from the logbooks reveal sailing courses of fishing vessels and pinpoint fishing locations in the past (see Figure 6.7). The high-resolution logbook data can also be used to calculate and display, for example, seasonal distributions of fish catches and effort during the 1850s (see Figure 6.8). Accordingly, the geographic positions of marine species sighted and recorded in the logbooks serve to improve knowledge of the extent and character of changes in fishing effort and, consequently, the populations and distributions of fish on the Scotian Shelf.

Discussion

Digitization of historic fishing grounds and logbook data using GIS software is a first step in assessing the accuracy and usefulness of historical records in reconstructing past fish abundances and distributions of marine species as well as fishing effort.

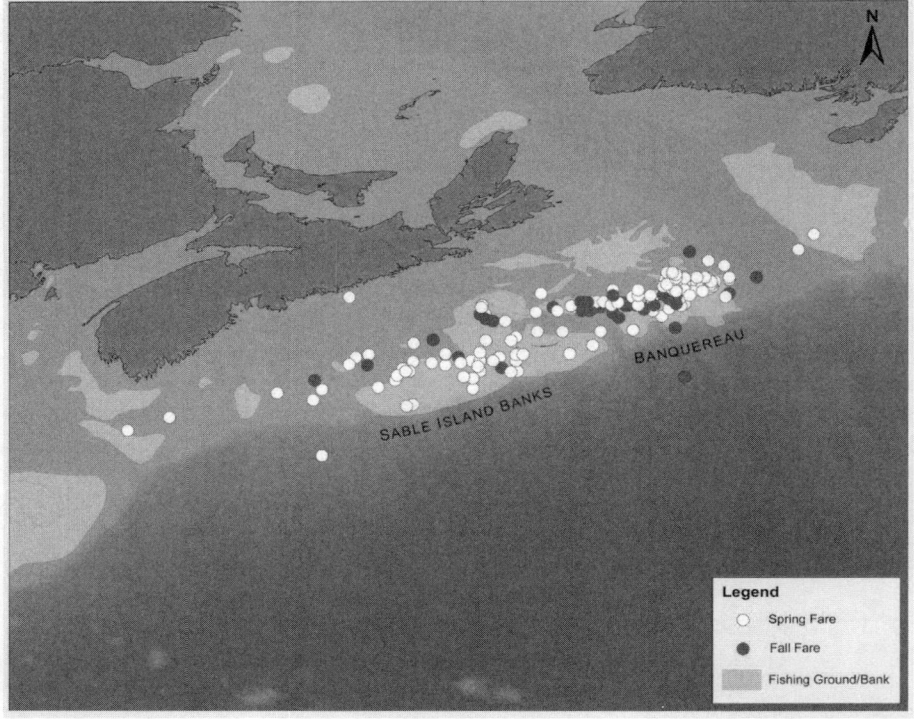

Figure 6.8 *Seasonal fishing by US vessels on the Scotian Shelf showing a concentration of effort on Banquereau and Sable Island Banks*

Source: *Fishing logbooks,* National Archives and Records Administration, Waltham, MA; view colour version at www.hull.ac.uk/oceanspast

Through identification of places of fishing effort, past abundances and distributions of marine species may be estimated locally and regionally. Additionally, how fishermen reacted or adapted to climate change, management or legislative actions, market demand and new fishing technologies can be measured. For example, the mapping and analysis of Beverly, MA fishing logbooks suggests that fishermen on the Scotian Shelf in the 1850s shifted their efforts from Banquereau to the Western or Sable Island Banks. This shift coincided with a drop in catch per unit effort (CPUE), indicating that the Scotian Shelf fishery was already in trouble by the 1850s (Rosenberg et al, 2005). Initial estimates reveal that the 19th-century cod population baseline on the Scotian Shelf was two orders of magnitude greater than today's population (Rosenberg et al, 2005). With the addition of site-specific data from 19th-century fishing logbooks and charts, such baseline estimates can now be keyed to specific regions or local environments.

Mapping fish catch records to fishing grounds or banks, rather than to where fish was landed, makes it possible to trend and estimate total removals of biomass for specific places, identify shifts in species composition and seasonality over time, and improve understanding of inter-bank, food-web and ecosystem linkages. When catch records are site-specific, shifts in the distribution of catch within a region can be analysed to yield relative measures of productivity and an understanding of the resilience of particular areas to fishing pressure. Moreover, place-based historical catch data can be used to identify minimum population or abundance estimates (based on total removals) and, when compared to modern biomass estimates, provide an assessment of the relative health of fish populations. However, critical to such analyses is identification of the temporal scale and resolution, and spatial precision and accuracy of historical datasets. Lacking definitive metadata, the use of historical data and analyses in fisheries and ecosystem-based management is problematic and may lead to potentially deleterious management decisions and negative policy actions. For example, blanket use of regional, aggregated historical datasets as proxies to infer recovery of sub-regional or localized diversity (e.g. Lotze et al, 2006) may direct management and policy to loosen fishing restrictions that can further degrade local ecosystem conditions. When the spatial and temporal limits of historical datasets are known and documented, linkages between fish catch and grounds can be defined accurately and precisely for local–regional comparative analyses of species composition, diversity, abundance and the behaviour or effort of fishers.

The results of this mapping project contribute directly to discussion of at least three marine conservation and management issues: first, identifying historically productive marine or essential fish habitats; second, rebuilding fish stocks and setting harvesting targets and thresholds; and third, data access and management. The National Research Council, which advises on the direction of science and policy in the US, has recently highlighted the need for historical research:

> *...that improves understanding about the extent of fishing effects on marine ecosystems and promotes the development of ecosystem, food-web, and species-interaction models and their incorporation into management decisions. ... Historical data assessments will be necessary to provide new insights about past species abundances and interactions. Comprehensive analyses of existing data can be applied to ongoing changes in target species, and they can help identify changes in habitat and non-target species that are thought to indicate ecosystem status. Landings data, narratives and descriptions, fisheries-independent data, phytoplankton and plankton records, satellite data, and archived specimens should all be considered when conducting these types of analyses. Examining these time-series or snapshot data in ecosystem and food-web models may provide the best approach to synthesizing long-term data and identifying alternative future scenarios to evaluate*

policy choices. Determining historical levels of exploited populations, and their natural fluctuations, should also provide a baseline around which to establish future management actions, including the setting of recovery goals. (NRC, 2006, pp94–95)

The Council also calls for improved data management strategies in order to 'increase access to historical data, incorporate data from disparate sources, and support the policy-making process' (NRC, 2006, p96). It justifies this call as follows:

Access to data from diverse sources will facilitate the transformation of these data into useful information that leads to model-based scenario analysis and informed decisions. Large-scale modeling of marine ecosystems requires facile integration of data from multiple sources. Better data management is fundamental to implementing ecosystem-based management of fisheries. (NRC, 2006, p96)

A place-based, GIS approach to managing historical fisheries and marine environmental data can provide an important visual and analytical tool for conservationists, ecologists, policy-makers and the public to interpret and know the past, current and future health of marine ecosystems and animal populations. Historical fishing charts provide significant baseline geographic information about the maritime landscape of the Gulf of Maine and for place-based mapping and analyses, but they are rare and of limited quantity. In order to improve analytical capabilities and understanding of past human impacts upon the marine environment, data mining and mapping of complementary qualitative and quantitative historical information from landing records, interviews with fishers, logbooks and ledgers, diaries and journals, market and newspaper reports, early scientific explorations, as well as natural history museum specimens and the zooarchaeological record is needed. Through the process of mapping historically productive Gulf of Maine fishing grounds and related catch data, critical marine habitats and conservation areas may be identified that are currently unknown or unrecognized because of a degraded condition. Moreover, spatial analyses may reveal correlations between fishing effort and local population extinctions, or how specific human behaviours or fishing technologies have affected fishery health. Finally, from historical analyses we may understand better the socio-cultural and economic consequences of marine resource depletion. As illustrated in this chapter, GIS maps and analyses of historical data can contribute directly to understanding the present condition or health of our marine ecosystems.

Acknowledgements

Special thanks are extended to GMCP historians Karen Alexander and William Leavenworth for their contributions to this research paper, and to supporting UNH faculty members and GMCP principal investigators, Andrew Rosenberg and Jeffrey Bolster. The archival sources for this project were obtained primarily from the National Archives Records Administration (NARA) Regional Repository at Waltham, Massachusetts, the James Duncan Phillips Library, the Peabody Essex Museum (PEM), Salem, Massachusetts, and the private collection of William Leavenworth. Financial support for the project was provided by the Alfred P. Sloan Foundation via the History of Marine Animal Population (HMAP) and the Census of Marine Life (CoML) programmes, the Mia J. Tegner Memorial Research Grants in Marine Environmental History and Historical Marine Ecology, the New Hampshire SeaGrant and the Hayes Chair of New Hampshire History at UNH.

References

Altham, E. (1623) 'Letter from Emmanuel Altham to Sir Edward Altham', September 1623, www.mayflowerhistory.com/PrimarySources/AlthamLetters.pdf, accessed 31 August 2007

Brereton, J. (1903) *A Briefe and True Relation of the Discouerie of the North Part of Virginia, by John Brereton; Reproduced in Facsimile from the First Edition of 1602, with an Introductory Note by Luther S. Livingston*, Dodd, Mead & Co., New York

Burrage, H. S. (ed) (1906) *Early English and French Voyages, Chiefly from Hakluyt 1534–1608*, C. Scribner, New York

Census of Marine Life (no date) *History of Marine Animal Populations: Understanding the Past is the Key to the Future*, www.cmrh.dk/hmaphandout.pdf

Goode, G. B. (1887) 'Section III: The fishing grounds of North America with forty-nine charts', in G. B. Goode (ed) *The Fisheries and Fishery Industries of the United States*, US Commission of Fish and Fisheries, Government Printing Office, Washington, DC

Grenier, R. (1988) 'Basque whalers in the New World: The Red Bay wrecks', in Bass, G. F. (ed) *Ships and Shipwrecks of the Americas*, Thames and Hudson, London, pp69–84

Haedrich, R. L. and Hamilton, L. C. (2000) 'The fall and future of Newfoundland's cod fishery', *Society and Natural Resources*, vol 13, pp359–372

Innis, H. (1940) *The Cod Fishery: The History of an International Economy*, Yale University Press, New Haven CT, p15

Lotze, H. K., Lenihan, H. S., Bourque, B. J., Bradbury, R. H., Cooke, R. G., Kay, M. C., Kidwell, S. M., Kirby, M. X., Peterson, C. H. and Jackson, J. B. C. (2006) 'Depletion, degradation, and recovery potential of estuaries and coastal seas', *Science*, vol 312, pp1806–1809

Myers, R. and Worm, B. (2003) 'Rapid worldwide depletion of predatory fish communities', *Nature*, vol 423, pp280–283

NRC (National Research Council) (2006) *Dynamic Changes in Marine Ecosystems: Fishing, Food Webs, and Future Options*, National Academy Press, Washington, DC

O'Leary, W. (1996) *The Maine Sea Fisheries: The Rise and Fall of a Native Industry, 1830–1890*, Northeast University Press, Boston, MA

Palfrey, J. G. and Palfrey, F. W. (1865) *History of New England, Volume 1*, Little, Brown, and Co., Boston, MA

Pauly, D. (1995) 'Anecdotes and the shifting baseline syndrome of fisheries', *Trends in Ecology and Evolution*, vol 10, no 10, p430

Pope, P. E. (2004) *Fish into Wine: The Newfoundland Plantation in the Seventeenth Century*, University of North Carolina Press, Chapel Hill, NC, p15

Rich, W. (1929) 'Appendix III to the Report of the US Commissioner of Fisheries for 1929', in *Fishing Grounds of the Gulf of Maine*, US Department of Commerce, Bureau of Fisheries Document No 1059, Government Printing Office, Washington, DC

Rosenberg, A. A., Bolster, W. J., Alexander, K. E., Leavenworth, W. B., Cooper, A. B. and McKenzie, M. G. (2005) 'The history of ocean resources: Modelling cod biomass using historical records', *Frontiers in Ecology and the Environment*, vol 3, pp84–90

7

There She Blew! Yankee Sperm Whaling Grounds, 1760–1920

John L. Bannister, Elizabeth A. Josephson, Randall R. Reeves and Tim D. Smith

Sperm whaling – hunting the largest of the toothed whales, the sperm whale (*Physeter macrocephalus*) – occurred historically in two major phases. Both phases used essentially the same technology as other whaling on the large baleen whales, such as right, blue, fin and humpback. The first phase, 'primitive' or 'open-boat' whaling, employing hand-held harpoons and lances, was only possible on the relatively slow swimming right, humpback and sperm whales. The second phase, 'modern whaling', using explosive-tipped harpoons mounted on fast catcher boats, allowed the additional pursuit of the faster swimming species, such as blue and fin whales.

While baleen whaling had been conducted since the 12th century at least, open-boat whaling on sperm whales occurred over only a relatively short period from the 18th to early 20th centuries; it was also more organized and systematic. Although other nations hunted whales, the American ('Yankee') sperm whale fishery was dominant and had a great effect on the fledgling economy of the US and the spread of American culture worldwide (Ellis, 2002).

In addition to the Americans, vessels from Britain, France, Germany, The Netherlands and Portugal were actively engaged in sperm whaling at various times. Whaling also took place from 'colonial' centres, particularly Australia and New Zealand. It was also recorded from Chile, and there are occasional records of whaling vessels originating from Denmark, Norway and what is now Canada (e.g. Nova Scotia and New Brunswick). However, as Broeze (1977) observes:

> *Even in the 1840s, when the greatest expansion in Pacific [sperm] whaling took place and almost all European maritime nations tried to*

> *establish their whaling industries, they were at no time able to match the Americans in daring, inventiveness or physical endurance.*

Open-boat whalers hunted sperm whales throughout the world's oceans, apart from the polar regions; but most of their catch came from relatively well-defined locations: the whaling grounds. Although the use of some grounds changed over time, their locations seem to have been relatively well defined in both space and time.

This chapter is designed to improve understanding of the spatial and temporal effects of Yankee whaling on sperm whale stocks (see Smith and Reeves, 2003). First, based on published information, the grounds are identified by name, where necessary reconciling multiple names; then their geographical boundaries are delineated. Their seasonal use is described, together with their progressive opening up and decline. The results are then compared with information from a subset of logbook extracts used in compiling 'Townsend's Charts' (Townsend, 1935). These charts show the positions of American whaling vessels on days when they caught sperm whales, worldwide, during the span of the open-boat fishery, from the mid 18th to the early 20th century. Finally, there are suggestions for further research into the general question of the reasons for decline in the fishery, the overall typicality of Townsend's charts and the factors responsible for the presence of sperm whales on particular grounds.

Methods and sources

This study had three specific aims: to identify the grounds by name and location; to evaluate those results against a sample of logbook extracts; and to establish seasonal and historical patterns of use. Seven main sources were used: Beale (1835, 1839); Wilkes (1845); Maury (particularly 1851, 1852); Scammon (1874); Clark (1887); Townsend (1935); and Gilmore (1959). Beale concentrates on the British fishery, while all the others are essentially concerned with the American (Yankee) industry. Maury and Townsend provide extensive detailed charts based on where whales were seen (Maury) or killed (Townsend). Gilmore compiles information from Townsend's charts, as well as from modern records, particularly on the use of grounds by season. Also available is a subset of data extracted from the original logbooks used in Townsend's study, termed here as 'Townsend extracts' (see Bannister and Mitchell, 1980). The available extracts are from logbooks of whaling vessels with names beginning with the letters A to J. The locations of whale ships on days when one or more sperm whales were killed were digitized by Josephson and used to prepare charts of the whaling grounds.

Notwithstanding its comprehensiveness, there are problems with the information provided by Maury. For example, he has records of right whales off the Australian west and east coasts well north of any shown by Townsend, without any obvious reason for the discrepancy. There are other anomalies, including a lack of sperm whales north of 40° N in the North Pacific, even though the charts show right whales there (Bannister, 1986). The general validity of Maury's charts has been challenged: one study compared them to the original data sheets, and determined that faulty extraction (from logbooks) and transcription (from data sheet to chart) confused right whales with sperm, 'finback' and possibly humpback whales (Reeves et al, 2004). Further concerns about Maury's charts have been raised elsewhere (see Chapter 8 of this book).

While Maury's information has not been used here, information from the other authors has been compared with that provided by Townsend. These include Beale, Wilkes, Scammon and, in particular, Clark. Townsend's sperm whale charts, which represent a catch of some 37,000 sperm whales between 1761 and 1920, have been used extensively. An example of part of one of his charts is shown in Figure 7.1.

WHERE THE GROUNDS WERE

Names

The published descriptions of places where whaling was conducted vary in number and coverage, particularly in their use of the term 'ground'. Three definitions were adopted for this study comprising 'ground', 'locality' and location'. 'Ground' is a colloquial name for a well-recognized whaling area. This encompasses four different types of spatial parameter. The first comprises 'strictly colloquial', using, for example, the name of the (presumed) discoverer or (presumed) first vessel to whale there, usually with no obvious local connection, such as *Commodore Morris*, *Carroll*, *Steen* or *Archer*. The 'strictly geographical' refers to areas such as the 'Coast of Japan', 'Galápagos' or 'On the [equatorial] Line'. Ground can also include areas based on coordinates, such as 'Twelve-Forty', 'Two Thirty-sixes'. And it can also involve broadly geographical places, including 'Offshore', 'Western', 'Southern' and 'Middle'.

'Locality' is a geographical area, not obviously colloquial, without the epithet 'ground' in the text (and often more extensive than 'Ground'). Examples of this include 'Canaries–African Coast', 'North of the Hawaiian Islands' and 'Almost across the equatorial Pacific'. The final definition is 'location', which is essentially a combination of 'ground' and 'locality', representing the whaling 'ground' used generically by many authors.

The published descriptions vary in number and coverage, and particularly in their use of 'Ground' (as defined; note upper case 'G'). Beale, for example, has limited overall coverage (but considerable detail for one ocean basin: the South Pacific) and records only one Ground, *sensu strictu*. Wilkes and Scammon identify some Grounds, as does Clark. However, Townsend is much more comprehensive in referring to Grounds rather than localities (Townsend, 1935). The authors use

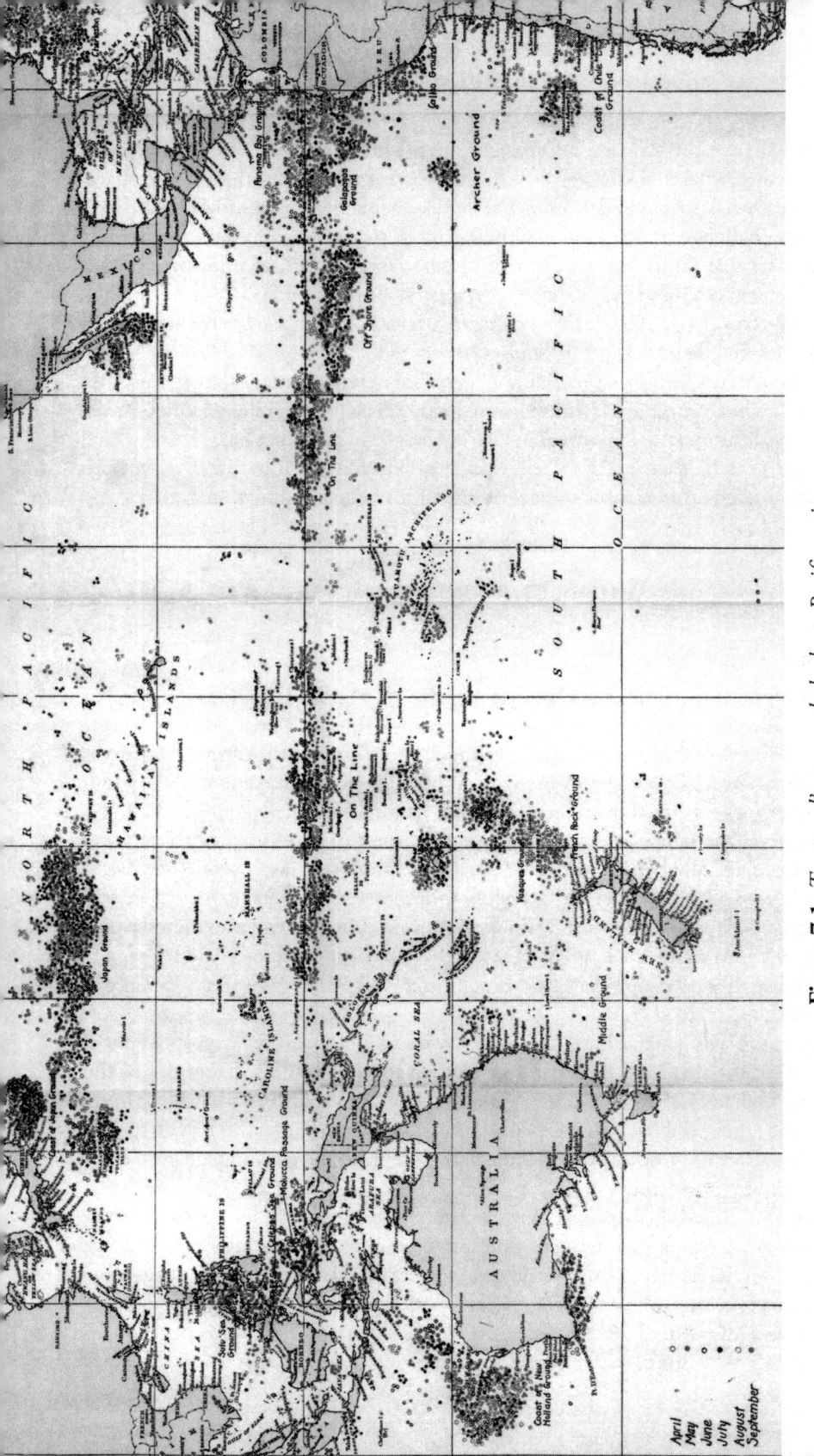

Figure 7.1 *Townsend's sperm whale chart, Pacific section*

Note: This extract shows positions of whale ships on days when they caught sperm whales for the six-month period of April to September.

Source: Townsend (1935); view colour version at www.hull.ac.uk/oceanspast

localities where Grounds are not named as such, sometimes – but not consistently – with coordinates. Table 7.1 summarizes the numbers of locations named by source for each ocean basin. It is worth noting the relatively small number of locations in Beale, Wilkes and Scammon (24, 25 and 33, respectively) compared with Clark (71) and Townsend (68). There is also a complete lack, or small number, of entries in Beale for the North and South Atlantic and North Pacific. This presumably reflects the fact that he was essentially dealing with the British fishery at the time of writing: 1835 to 1839. A low number of entries also apply to Wilkes, Scammon and Beale with regard to the South Atlantic. Conversely, there is a relatively large number of entries in Beale for the South Pacific. While some differences can be observed between Wilkes and Scammon, particularly in the North Pacific, there are similar numbers elsewhere. Despite their similar overall totals (71, 68), there is a marked difference between Clark and Townsend in their sub-totals for the North Atlantic (24, 16, respectively) and Indian Ocean (8, 14, respectively). Finally, there is a significant margin of difference between Clark and Townsend in number of Grounds recorded *sensu strictu*, with Townsend having virtually three times as many (44) as Clark (15).

Geographical boundaries

The various authors' inconsistency in the use of coordinates, both for Grounds and localities, has already been mentioned. For example, for the 25 North Atlantic locations recorded in his text, Clark gives complete rectangular (i.e. four) coordinates for 11, incomplete (three or two) for 5, and none for the remaining 9. Despite the fact that Townsend provides coordinates fairly readily in a preliminary

Table 7.1 *Numbers of sperm whaling locations identified by ocean basin and author*

Ocean	Source														
	Beale (1835, 1839)			Wilkes (1845)			Scammon (1874)			Clark (1887)			Townsend (1931, 1935)		
	G	L	T	G	L	T	G	L	T	G	L	T	G	L	T
N Atlantic	0	0	0	0	5	5	1	6	7	7	17	24	12	4	16
S Atlantic	0	1	1	1	1	2	1	1	2	2	6	8	7	3	10
N Pacific	0	5	5	0	5	5	0	9	9	2	8	10	7	3	10
S Pacific	1	10	11	2	6	8	3	6	9	4	17	21	10	8	18
Indian	0	7	7	0	5	5	0	6	6	0	8	8	8	6	14
Total	1	23	24	3	22	25	5	28	33	15	56	71	44	24	68

Notes: G = Ground; L = locality; T = total.

Source: see text

account (Townsend, 1931), they are given in only a few cases in his major work (Townsend, 1935). Coordinates have therefore been obtained directly from the plots on Townsend's individual 1935 charts, using subjective judgement for concentrations and allowing for outliers. For most areas, Townsend provides plots for the two six-monthly periods, April to September and October to March; but in this chapter a combined area has been adopted for the 12-month period.

A common criticism of Townsend's charts is that the plots extend well beyond the areas where whaling actually took place. This is certainly the case for humpback whales. A careful look at the charts shows the plots rarely overlapping; the result is that some grounds appear quite extensive even when centred on a specific bathymetric or oceanic feature, such as a seamount or shoal (Cawthorn, 1995). But Townsend (1931, p178) himself pre-empts such criticism in his 1931 paper:

> *It should be explained that on a few whaling grounds where the plottings are very dense, the draughtsman extended the areas slightly for lack of space. These are off the west coast of South Africa and off the east coast of South America; around the Galapagos Islands; off northern New Zealand and off the north coast of Japan. ... In other charts now in preparation [for the 1935 version], this difficulty is avoided by omitting some of the data rather than extend beyond actual whaling limits.*

The 'crowding' effect may well be more of a problem with coastal species, such as humpbacks, than with more offshore species, including sperm whales. For present purposes, the main concentration, as judged from the plottings, has therefore been regarded as a true reflection of the whaling location in question (but see Conclusions, below).

It is not always easy to determine whether one defined location is the same as another. Beale's numerous locations for the North and South Pacific cannot always be readily equated with present-day names (*The Times 1900 World Atlas* was invaluable in that context). However, in some cases, the same or very similar words and coordinates are used by successive authors as their predecessors. This seems particularly the case for Scammon and Wilkes. For example, Scammon has 'Azores/Western Islands' compared with Wilkes's 'Off the Azores or Western Islands'. Both, for a major ground off Peru, have '*Offshore* 5–10° S, 90–120° W'. On the other hand, there are only two cases (*Carroll*, in the South Atlantic, and *Middle*, in the South Pacific) where different authors seem to use the same name for different locations.

There are, nevertheless, discrepancies in geographical definitions, usually in terms of extent rather than geographical position, one large area sometimes encompassing one or more smaller, or two or more, areas overlapping. The major comparison is again between Clark and Townsend. For that comparison, the positions of the 68 localities whose coordinates had been extracted from Townsend's charts were plotted. These were then compared with plots of localities for which

Clark gives either coordinates or sufficiently detailed geographical descriptions to allow the locality to be pinpointed. Of the 71 localities thus listed by Clark, coordinates or sufficient equivalents are available in his text for 45, amounting to some 60 per cent.

From that comparison, 64 locations were identified. They comprise 39 Grounds and 25 localities (as defined). They are listed by number in Table 7.2 and by name and locality in Table 7.3. A geographic representation is illustrated in Figure 7.2a and Figure 7.2b. Accordingly, the tables and the figure represent a consolidation of locations used worldwide by Yankee sperm whalers during the 18th and 19th centuries.

WHEN THE GROUNDS WERE VISITED

Some authors, including Wilkes, Clark and Townsend, give information on the seasons and/or months when particular whaling areas were frequented. Townsend gives limited information in his text, but the plottings on his charts are individually coloured by month. Information from Townsend's charts was used in a valuable compilation by Gilmore (1959). He showed the world distribution of sperm whales based on both the charts and modern records. Of particular interest here are his classifications of whaling locations by season, defined as 'northern summer and southern winter', 'southern summer and northern winter', 'both northern and southern summer and winter', and 'year around'. His figure (with seasonal designations shaded) is reproduced here as Figure 7.3.

A compilation of seasonal information from the three main sources – Wilkes, Clark and Townsend – together with Gilmore's results (using his Townsend-derived material) is provided in Table 7.4. It shows general agreement among the various authors, but there are some anomalies. Among the most important – with numbers designated in Table 7.3 – is number 15 (*Twelve-Forty* Ground, North

Table 7.2 *Numbers of Yankee sperm whaling grounds and locations identified by ocean basin*

Ocean	Grounds	Localities	Total
N Atlantic	11	6	17
S Atlantic	7	3	10
N Pacific	4	6	10
S Pacific	10	7	17
Indian	7	3	10
Total	39	25	64

Source: see text

Table 7.3 Consolidated list of Yankee sperm whaling locations

Ocean	No	Name(s)[1]	Coordinates[2]	Comments[3]
North Atlantic	1	Western	27°–41°N, 34°–58°W	Clear distinction in Townsend between Nos 1 and 2 (adjacent to each other); blurred in Clark (esp by C16), C16, C32 and C19 extend combined area north beyond coordinates
		(the *Two Forties*, the *Two Thirty-sixes*)		
	2	Western Islands (= Azores)	35°–41°N, 23°–34°W	
	3	Southern	34°–41°N, 60°–74°W	
	4	Hatteras (incl *The Bahamas*)	29°–37°N, 68°–76°W	C13 (called *Hatteras*) is a point location only, and very much a western outlier. C11 extends No 5 to the west. Nos 3, 4, and 5 all overlap to some extent.
	5	Charleston	(a) 28°–33°N, 67°–76°W	
			(b) 29°–32°N, 74°–79°W	
	6	Commodore Morris	(a) 48°–51°N, 20°–24°W	C21 further NW than coordinates
			(b) 52°–54°N, 23°–25°W	
	7	The Shoals	46°–47°N, 52°–54°W	Small NW outlier of *Western* (No 1). Townsend only, but see No 1
	8	Steen	31°–35°N, 19°–25°W	Southeast extension of *Western Is* (No 2)
	9a	S Portugal/Spain (= C St Vincent–Gibraltar)	36°–37°N, 06°–09°W[4]	Clark only
	9b	Coast of Morocco	32°–37°N, 06°–11°W,	Townsend only, summer only
	10	Unnamed (north of Madeira)	34°–38°N, 15°–19°W	Northeast extension of, and could be combined with, *Steen* (No 8). Townsend only
	11	Canaries–African Coast	26°–30°N, 11°–19°W	
	12	San Antonio (the *Twenty Twenties*) (= C Verdes)	14°–22°N, 19°–29°W	
	13	Cornell	01°–09°N, 18°–28°W	
	14	Gulf of Guinea (near Fernando Po I)	3°30′N, 8°40′E[5]	Clark only
	15	The Twelve-Forty	10°–17°N, 35°–46°W	
	16	Gulf of Mexico	26°–29°N, 86°–90°W	
	17	Caribbean Sea/W Indies: a) Haiti, b) Anguilla/Antigua	16°–21°N, 60°–75°W	
South Atlantic	18	Coast of Brazil	02°–28°S, 30°–42°W	
	19	Platte	26°–40°S, 41°–56°W	
	20	False Banks	42°–48°S, 53°–59°W	
	21	Walvis (Walfisch) Bay	03°–23°S, 03°–13°E	
	22	Coast of Africa	06°–10°S, 09°–15°W	Townsend only; included in *Coast of Africa*, No 22
	23	Around Ascension	26°–29°N, 86°–90°W	
	24	Pigeon	28°–37°S, 17°–30°W	Townsend only
	25	Carroll (incl St Helena)	13°–18°S, 07°W–02°E	C48 (a line only) extends No 26 to the east
	26	Tristan	29°–39°S, 01°–14°W	C49 (a point only) is a northern outlier of T26
	27	Tristan–C of Good Hope	33°–38°S, 03°–12°E	
	28	High Latitudes (50–60°N)		Scammon only. Not plotted
North Pacific	29	Japan	33°–40°N, 150°E–175°W	Beale: '*Japan offshore*'. Clark and Townsend's grounds extend west to *Coast of Japan*, No 30; Clark's extends to the east (to Hawaiian Is, No 32), in line with Scammon ('across the N Pacific between 27 and 35°N')
	30	Coast of Japan	31°–38°N, 141°–150°E	Nos 30 and 31 could be considered together
	31	Japanese Coast to the Bonin Islands (incl Bashee Is [Bataan])	24°–31°N, 136°–150°E	
	32	[Near the] Hawaiian Islands	15°–34°N, 149°–168°W	
	33	Northwest Coast	55°N, 144°W	Point location only. Not in Clark. Townsend: 'only as stragglers'
	34	California Coast	20°–28°N, 109°–118°W	
	35	Panama Bay	04°–08°N, 78°–90°W	Clark has it to 90°W from the coast, and extending south of the equator, along the coast, to *Archer* (his C53)
	36	South west of Clipperton I	05°–09°N, 110°–116°W	C38 is at ca 110°–120°W.
	37	[East Indies] (incl China Sea/Sulu Sea/Celebes Sea/Molucca Passage)	(a) 04°–12°N, 115°–122°E	Considered together: in Townsend, 4 elements ± overlapping, plus an extension on the NW coast of New Guinea; also outliers towards the Gulf of Siam
			(b) 05°N–03°S, 118°–126°E	
			(c) 02°S–04°N, 124°–130°E	
			(d) 00°–02°N, 130°–136°E	

Region	No	Name	Coordinates	Notes
South Pacific	38	*Galápagos*	04°N–17°S, 78°–98°W	T37 much larger than C57; embraces part C54 (his *Panama Bay*, see No 35) and whole of C56 (his *Callao*, see No 52)
	39	*Offshore*	02°N–08°S, 97°–118°W	T38 larger than C58(a) (thin strip). C58(b) extends T38 to south and west. T38 just overlaps with No 38 (*Galápagos*) to east. All overlap with No 40 (*On the Line* – E) in the west
	40	*On the Line* [East]	03°N–05°S, 112°–139°W	C59 and T39 both overlap with No 39 (*Offshore*) in the east
	41	Almost across the Equatorial Pacific [= *On the Line* [West]]	03°N–06°S, 162°E–154°W	
	42	[S Coral Sea]	21°–26°S, 154°–160°E	Subdivided as per Townsend's two main plots (T42/44, 43)
	43	Around New Guinea (incl New Ireland, New Britain, Solomon Is, N Coral Sea, New Caledonia)	(a) 01°N–04°S, 140°–145°E (b) 00°–10°S, 149°–154°E	
	44	[Fiji–Samoa–Tonga]	14°–20°S, 174°E–178°W	C62 extends well to the east to include T51 (No 49)
	45	*Middle*	29°–38°S, 156°–168°E	By definition, between Aust and NZ
	46	New Zealand; *Vasquez; French Rock*	(a) 22°–36°S, 166°E–171°W (b) 35°–38°S, 164°–166°W	Complex area covered by T48 (summer). Slight overlap with No 45 (*Middle*) in SW C64, 65 (= No 46(b)) are outliers to SE, not appearing on Townsend plot
	47	Chatham Is	42°–46°S, 172°–179°W	
	48	South of New Zealand (= *Solander*)	46°–49°S, 165°–170°E	
	49	[Tuamotus, Marquesas, Society Is (Tahiti)]	07°–20°S, 140°–153°W	
	50	Coast of Chile	(a) 33°–46°S, 73°–79°W (b) 30°–38°S, 77°–87°W	Townsend plots have two separate elements: (a) coast of Chile, (b) Juan Fernández
	51	*Archer*	17°–21°S, 88°–93°W	Small southern extension of No 38 (*Galápagos*)
	52	*Callao*	08°–20°S, 70°–89°W	Overlaps extensively in west with No 38 (*Galápagos*), and slightly with No 51 (*Archer*). Nos 38 (*Galápagos*), 51 (*Archer*) and 52 (*Callao*) could be considered together
	53	Across the S Pacific between 21° and 27°S		Not in Townsend or Clark. Not plotted
	54	Cape Horn		Not in Townsend. Not plotted
Indian	55	Coast of Arabia (incl *Red Sea, Persian Gulf*)	14°–18°N, 50°–58°E	Persian Gulf (B19) only in Beale
	56	Unnamed [Off Somaliland]	03°–07°N, 53°–58°E	Small, offshore, only in Townsend
	57	Zanzibar	02°N–11°S, 39°–48°E	
	58	Mahe Banks[6]	0°–11°S, 50°–62°E	
	59	Delagoa Bay/Mozambique Channel/[South Madagascar]	(a) 11°–25°S, 35°–45°E (b) 23°–29°S, 32°–39°E (c) 25°–28°S, 43°–46°E	The three Townsend grounds form a ± natural grouping. T59 and 61 abutting C1 to its south; also coastwise SW towards Port Elizabeth
	60	East of Madágascar	(a) 12°–27°S, 47°–54°E (b) 17°–24°S, 54°–65°E	C2 joins No 60 with No 58 to the north and No 59 to the west
	61	South of Ceylon[7]	03°–07°N, 75°–80°E	
	62	Andaman Islands to Sumatra	(a) 04°–11°N, 91°–95°E (b) 10°–14°S, 105°–119°E	2 main elements, widely separated: T64a (Andaman Is); T64·b, s of Sumatra. Neither in Clark
	63	Coast of New Holland	(a) 10°–20°S, 108°–119°E (b) 20°–31°S, 102°–112°E (c) 33°–38°S, 111°–128°E	T65 in three elements – T65c a southern outlier. C71 much less extensive.
	64	Nr Crozet Is/St Paul/Kerguelen	(a) 45°S, 50°E (b) 38°S, 77°E (c) 49°S, 69°E	Three widely separated point locations, only in Townsend. Also scattered small groupings between Madagascar and (a) and (b), and eastwards. Scammon has 'High latitudes to 60°S'.

Notes: 1 Italicised names are 'Grounds' as recognized by the various sources. 2 Degrees of longitude and latitude, indicating ranges of approximate maximum rectangular coverage (or, in some cases, single point locations), based mainly on Townsend. 3 The annotations 'C' and 'T' refer to information in Clark and Townsend, as tabulated by the first author. 4 Cape St Vincent–Gibraltar. 5 Fernando Po I. 6 Wray and Martin (1983) include Saya de Malha (east–northeast of Madagascar, on the southeast–northeast borders of numbers 58/60). 7 Wray and Martin (1983) include Chagos Archipelago, to the southwest.

Source: see text

118 OCEANS PAST

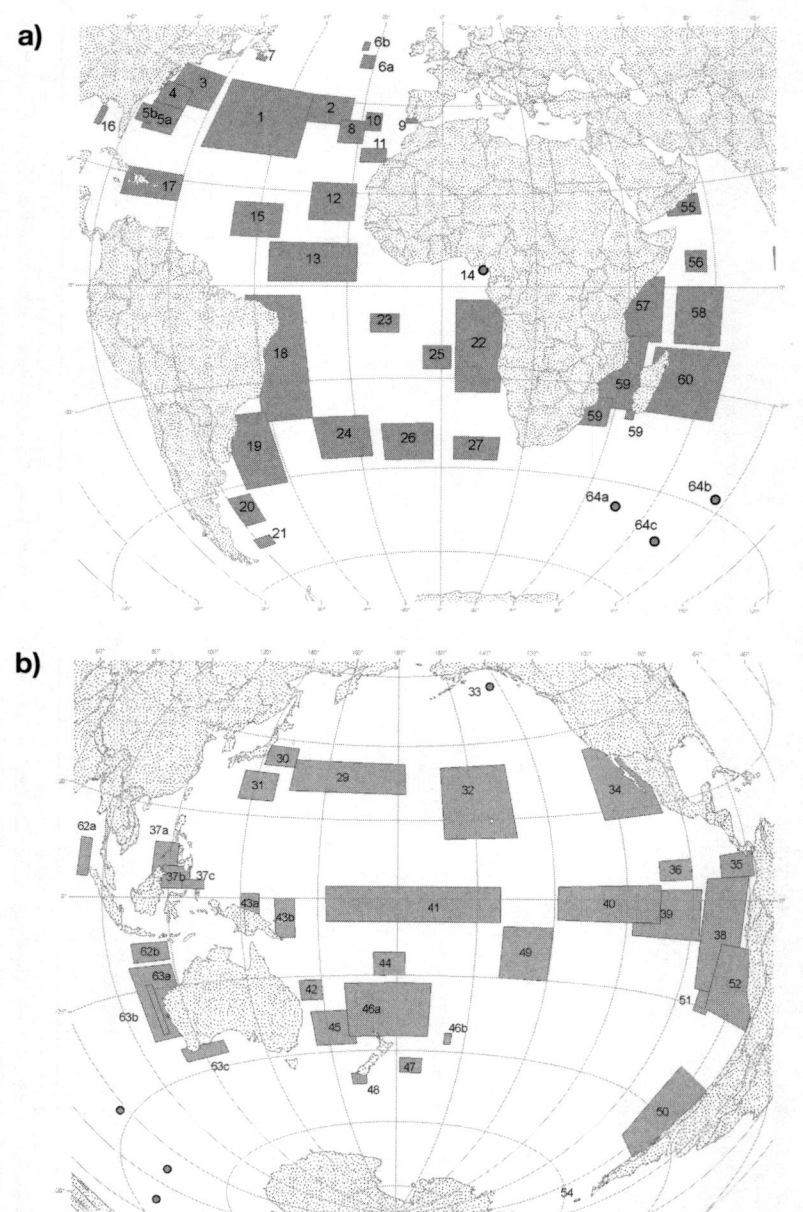

Figure 7.2 *Consolidated sperm whaling locations in (a) the Atlantic and Indian Oceans and (b) the Pacific Ocean*

Notes: Locations numbered as in Table 7.3.

Source: see Table 7.3; view colour version at www.hull.ac.uk/oceanspast

THERE SHE BLEW! YANKEE SPERM WHALING GROUNDS, 1760–1920   119

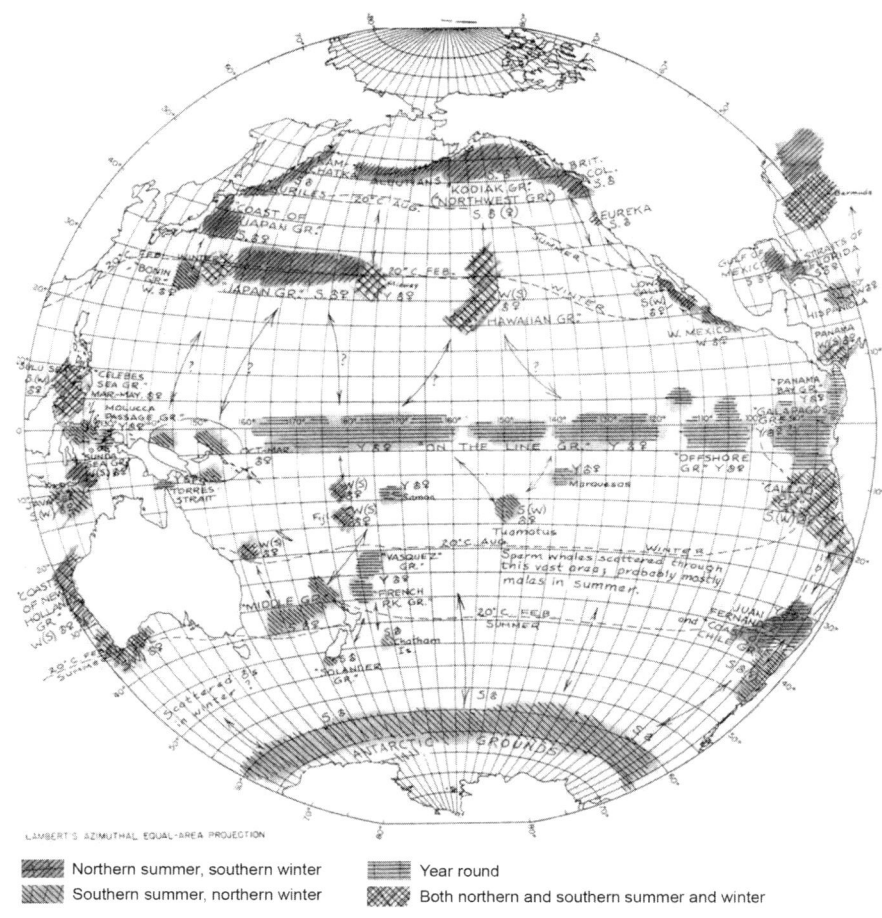

Figure 7.3 *Seasonal distribution of sperm whaling grounds in the Pacific Ocean*

Source: adapted from Gilmore (1959); view colour version at www.hull.ac.uk/oceanspast

Atlantic). This was stated by Clark to be frequented in March to May, and by Townsend in February to May; but Gilmore shows this to be a winter ground. In the case of number 31 (Japan Coast–Bonin Island) Wilkes specifies effort in July; Clark over May to October and November; Townsend between May and August; and Gilmore over the winter period. Wilkes has number 34 (California Coast) as frequented between November and January; Clark cites it as winter; Townsend, the May to January period; and Gilmore as over the summer. With regard to number 45 (*Middle* Ground, South Pacific), Wilkes describes whaling as taking place 'after May', whereas Townsend specifies December to March and Gilmore, once again, cites summer. While Wilkes also claims that number 49 (Tuamotus,

Table 7.4 *Seasonality of visitation, by location and source*

Ocean	No	Name(s)	Wilkes (1845) s	w	c	Clark (1887) s	w	c	Townsend (1931, 1935) s	w	c	Gilmore (1959) s	w	c
North Atlantic	1	Western (the *Two Forties*, the *Two Thirty-sixes*)				s–a, often into Dec			Apr–Sept			✓		
	2	Western Islands (= Azores)	s–Oct						Apr–Sept					✓
	3	Southern	May						Apr–Sept	Oct–Jan				✓
	4	*Hatteras* (incl *The Bahamas*)												
	5	*Charleston*					w–spg							
	6	*Commodore Morris*												
	7	*The Shoals*					a							
	8	*Steen*				Aug–Nov			✓					
	9	S Portugal/Spain (= C St Vincent–Gibraltar)							✓					
	10	Unnamed (NE of Madeira)												
	11	Canaries–African Coast					a						✓	
	12	San Antonio (the *Twenty Twenties*) (= C Verdes)		?Oct			w			a			✓	
	13	*Cornell*					w							
	14	Gulf of Guinea												
	15	*The Twelve-Forty*				Mar–May	w–early spg		Feb–May					
	16	Gulf of Mexico					w–early spg		Feb–May	✓				
	17	Caribbean Sea/W Indies a) Haiti, b) Anguilla/Antigua	Oct–Dec											
South Atlantic	18	Coast of Brazil		Jan–April					Oct–Mar/Apr			✓		✓
	19	*Platte*	Jan–Feb			Sept–May								
	20	*False Banks*				Nov–May								
	21	Unnamed (between Brazil Banks and Falklands)						✓						
	22	Coast of Africa (incl Woolwich [Walfisch] Bay)						✓	Dec–Mar[2]		All seasons	✓		✓
	23	Around Ascension												
	24	*Pigeon*										✓		✓
	25	*Carroll* (incl St Helena)										✓		✓
	26	*Tristan*		Mar–May										
	27	Tristan – C of Good Hope				Sept–May			Oct–Jan					
North Pacific	28	High Latitudes (50–60°N)							s–a					
	29	*Japan*	May–Nov			May–Oct	Nov		May–July			✓[3]		
	30	Coast of Japan	May–Nov			May–Oct	Nov		May–Aug			✓[3]		
	31	Between the Japanese Coast and the Bonin Islands (incl Bashee Is [Bataan])	July			May–Oct	Nov							
	32	[Near the] Hawaiian Islands	June–Oct						May–Aug cont to Jan					
	33	*Northwest coast*	Aug–Sept				✓		Apr–July			✓		

	#	Ground							
	34	California Coast			Nov–Jan	✓		May–Aug cont to Jan	✓✓✓
	35	*Panama Bay*							✓
	36	Further west in the ocean [than *Panama Bay*]							✓✓✓
	37	[East Indies] (incl *China Sea/Sulu Sea/Celebes Sea/Molucca Passage*)							
	38	*Galápagos*	Nov–Feb			✓			
	39	Offshore	Nov–Feb						
	40	On the Line [East]		July–Aug					
	41	Almost across the Equatorial Pacific [= *On the Line* [West]]				✓			
	42	[S Coral Sea]							✓✓
	43	Around New Guinea (incl New Ireland, New Britain, Solomon Is, N Coral Sea, New Caledonia)							
	44	[Fiji–Samoa–Tonga]		July–Aug					✓
	45	Middle		After May					✓
	46	*New Zealand*; *Vasquez*; *French Rock*	Mar–May					Dec–Mar	
	47	Chatham Is						Dec–May	
	48	South of New Zealand (= *Solander*)							✓✓
	49	[Tuamotus, Marquesas, Society Is (Tahiti)]		After May				Dec–Mar mainly	✓
South Pacific	50	*Coast of Chile* (= *Onshore*)		Sept–May					
	51	*Archer*					'sometimes'		
	52	*Callao*					✓	Dec–Mar mainly	
	53	Across the S Pacific between 21° and 27°S							✓
	54	Cape Horn							
	55	*Coast of Arabia* (incl *Red Sea*, *Persian Gulf*)		Oct–April					✓✓✓
	56	Unnamed [Off Somaliland]							✓✓✓
	57	*Zanzibar*							
	58	Mahe Banks[4]							
	59	*Delagoa Bay/Mozambique Channel/*[South Madagascar]	Mar–Apr	May–July					
	60	*East of Madagascar*							✓
	61	*South of Ceylon*[5]							✓✓
Indian	62	Andaman Islands to Sumatra							✓✓
	63	*Coast of New Holland*		Oct–May[6]	Mar–July[7]				✓
	64	nr Crozet Is/ St Paul/Kerguelen							✓

Notes: 1 s = summer; w = winter; c = combined; spg = spring; a = autumn/fall. 2 Woolwich Bay. 3 Richards (2000, p190) says: "Townsend's charts show all three grounds [numbers 29, 30 and 31] were highly seasonal, with most whales taken in June and July.' 4 Wray and Martin (1983) include Saya de Malha (east-northeast of Madagascar, on the southeast–northeast borders of Nos 58/60). 5 Wray and Martin (1983) include Chagos Archipelago, to the southwest, as does Clark (but as 'very doubtful ground … not often been frequented.… The proper season is the southwest monsoon'). 6 'near the land'. 7 'offshore'.

Source: see text

Marquesas, Society Island) is mostly 'after May', Gilmore describes the main period as summer. Lastly, with regard to number 50 (*Coast of Chile/Onshore* Ground), Clark places the focus of whaling as September through to May and 'sometimes' in winter, while Townsend regards it as 'mainly' December to March and Gilmore specifies winter.

Information from various authors – including those already quoted, but also others, particularly Wray and Martin (1983) for the western Indian Ocean, Richards (1998) and Bannister et al (1981) for the North Pacific, and Richards (1982) for the South Pacific – is mostly consistent with Gilmore's overall conclusions. Thus, in Figure 7.3, most Northern Hemisphere locations north of 25° N, and most in the Southern Hemisphere south of 25° S are, as might be expected, summer grounds. Conversely, those few locations just south of 25° N in the Northern Hemisphere and just north of 25° S in the Southern Hemisphere are generally winter grounds. Most designated by Gilmore as 'year around' grounds occur near the equator, especially in the Pacific (but rather further south in the South Atlantic and Indian Oceans). Those designated 'both northern and southern summer and winter' occur between there and about 30° N and 30° S. The more important of the latter are: in the North Atlantic, *Charleston, Hatteras* (numbers 4 and 5) and *Western Islands* (number 2); in the South Atlantic, *Coast of Brazil, Platte, False Banks* (numbers 18, 19 and 20); in the North Pacific, *near the Hawaiian Islands* (number 32); in the South Pacific, *Callao* (number 52); and in the Indian Ocean, *Coast of New Holland* (number 63).

OPENING UP, MAJOR USE AND DECLINE

Information from the literature

Yankee sperm whaling had its small beginnings off New England early in the 18th century. Clark refers to the first capture in about 1720, with a fishery between Cape Cod and Bermuda from June to August (Clark, 1887, quoting Beale). This was followed by a gradual spread throughout the Atlantic until operations moved into the Pacific, around Cape Horn from 1786, and into the North Pacific by 1818. Whaling had begun in the Indian Ocean, presumably via the Cape of Good Hope, by at least 1791. The great prosperity of the American sperm whale fishery climaxed in 1837, based on the 'exceedingly profitable' Pacific grounds (Clark, 1887, p10). However, by the time of Clark's publication in the late 1880s, whaling had ceased in many formerly lucrative areas, and others were by then seldom used.

Whaling in the Atlantic was one exception to the above. There it might be assumed that the early activity, and that ocean's constant traversing *en route* elsewhere, would have reduced its whale stocks to the point where whaling would no longer have been profitable by about the 1870s. However, Starbuck (1964, p110) states otherwise:

> ... [sperm whaling] was practically abandoned ... after it became necessary to make voyages of four, five and even six years, and then seldom return with a full cargo.... Always excepting, of course, Atlantic whalers.... [In that ocean, whaling] has always been pursued by the bulk of the Provincetown whalers and by quite a fleet of schooners and brigs from other ports. There is an occasional revival ... in larger vessels at intervals of a few years, at present [circa 1875] some of the most successful voyages being made by ships and barks cruising for sperm whales in this ocean.

But later Starbuck tempers that with 'the persistency with which any field is followed up ... makes its yield at least but temporary' (Starbuck, 1964, p113).

The overall picture by location within ocean basin is given in Table 7.5. There, each 'x' entry represents a specific date mentioned by a particular author. Those designated 'c' represent the 'present grounds' – taken as circa 1880, as indicated in Clark's 1887 chart (plate 183). The lines joining each entry are assumed, but indicate the length of time an area was probably extant.

The overall decline of the sperm whale fishery from the mid 19th century is mentioned repeatedly in the literature. Clark (1887, p7) emphasizes it as follows:

> Many of the grounds where vessels were formerly very successful are now entirely abandoned and others but seldom visited. There are now no sperm whalers from the United States on the Indian Ocean or North Pacific grounds, and very few cruising in the west Pacific Ocean, but nearly all of the vessels at present engaged in this branch of the fishery resort to the grounds in the North and South Atlantic and the eastern part of the South Pacific.

There is, however, an anomaly between Clark's text and the 'present grounds' depicted on his 1887 chart. Of the 53 grounds in Table 7.5, 37 (70 per cent) are recorded (largely on the basis of Clark's chart, shown as 'c' in Table 7.5) as still or probably extant around 1880. From Table 7.5, North Pacific whaling was clearly considerably reduced by around 1880, with only two of the grounds still in use (*Japan–Bonin Island*, number 31, and *Unnamed, Further West* [than *Panama*], number 36). Clark's chart also shows some whaling still on many other grounds. Yet many of these – at least as plotted on his chart – were considerably reduced by comparison with those recorded on his chart as 'abandoned', and most of those were in the Pacific and Indian Oceans. But in the Atlantic, only 3 of the 18 on his chart are shown as reduced by around 1880. This accords with his own conclusions quoted above, as well as with Starbuck's comments about the continued profitability of whaling in the Atlantic until the 1870s, at least.

Table 7.5 *Main periods of occupation of grounds by ocean basin*

Ocean	Ground	1700	1710	1720	1730	1740	1750	1760	1770	1780	1790	1800	1810	1820	1830	1840	1850	1860	1870	1880	1890	1900
North Atlantic	No 4. *Hatteras*	x							x					x	x	x				c		
	No 1. *Western*							x.................................x											c		
	No 7. *The Shoals*							x_x_x														
	No 14. *Gulf of Guinea*							x					x									
	No 2. *Western Islands*							x_					x_x_x_x							c		
	No 3. *Southern*								x													
	No 11. *Canaries–African Coast*								x_				x									
	No 5. *Charleston*								x					x						c		
	No 12. *San Antonio*								x							x				c		
	No 17. *Caribbean Sea etc*								x_				x_			x				c		
	No 16. *Gulf of Mexico*															x				c		
	No 6. *Commodore Morris*																	x...........x		c		
	No 8. *Steen*																			c		
	No 9. *S Portugal/Spain*																			c		
	No 13. *Cornell*																			c		
	No 15. *Twelve-Forty*																			c		
South Atlantic	No 18. *Coast of Brazil*								x_x							x_				c		
	No 21. *Brazil Banks–Falklands*								x_x													
	No 22. *Coast of Africa*									x...........x										c		
	No 20. *False Banks*									x												
	No 19. *Platte*									x...........x										c		
	No 25. *Carroll*															x_				c		
	No 23. *Around Ascension*																			c		
	No 24. *Pigeon*																			c		
	No 27. *Tristan–Cape of Good Hope*																			c		
South Pacific	No 54. *Cape Horn*									x.......................................x									c		
	No 50. *Onshore*											x_x_x				x_				c		

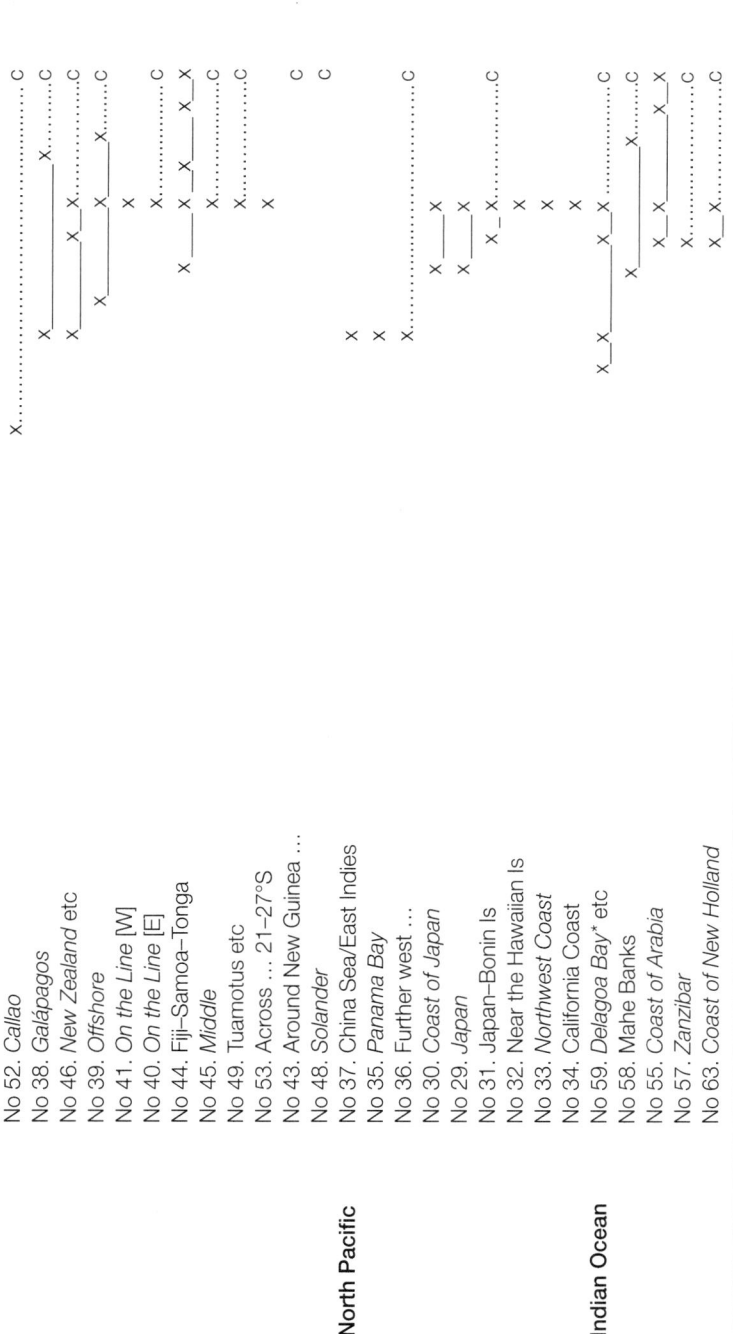

No 52. *Callao*
No 38. *Galápagos*
No 46. *New Zealand* etc
No 39. *Offshore*
No 41. *On the Line* [W]
No 40. *On the Line* [E]
No 44. *Fiji–Samoa–Tonga*
No 45. *Middle*
No 49. *Tuamotus* etc
No 53. *Across … 21–27°S*
No 43. *Around New Guinea …*
No 48. *Solander*

North Pacific
No 37. *China Sea/East Indies*
No 35. *Panama Bay*
No 36. *Further west …*
No 30. *Coast of Japan*
No 29. *Japan*
No 31. *Japan–Bonin Is*
No 32. *Near the Hawaiian Is*
No 33. *Northwest Coast*
No 34. *California Coast*

Indian Ocean
No 59. *Delagoa Bay** etc
No 58. *Mahe Banks*
No 55. *Coast of Arabia*
No 57. *Zanzibar*
No 63. *Coast of New Holland*

Notes: x = dates mentioned in the source text. c = shown as a 'present' (i.e. around 1880) ground in Clark's (1887) chart. A solid line between years indicates firm information on the length of time a ground was extant; dotted lines are as inferred from Clark's chart.

* Delagoa Bay itself was extant until about the 1840s, but other nearby areas ('Madagascar') continued throughout the 19th Century.

Source: see text

Information from 'Townsend extracts'

The extracts, available only for Yankee whaling vessels with names beginning with the letters A to J, are derived from 800 logbooks covering the period 1790 to 1922. In the present study, given that the information in the extracts has dates associated with the whale records, it has been used to illustrate the changing temporal and spatial nature of the sperm whale fishery, in confirmation or otherwise of the information given already.

The data were plotted by ocean basin and by four periods: 1790 to 1824; 1825 to 1849; 1850 to 1874; and 1875 to 1922. An example, showing the North Pacific over the whole period, is given in Figure 7.4. The full range of plottings by ocean basin and groups of years (despite the very few entries during the period of 1790 to 1824) shows that at the beginning of the period, whaling had already moved out of the Atlantic. Vessels had spread west around Cape Horn into the eastern South Pacific along the west coast of South America (*Coast of Chile/Peru*, or *Onshore* Ground, number 50), as well as east around the Cape of Good Hope into the southwest Indian Ocean. From 1825 to 1849 – the peak of the sperm whale fishery worldwide – whaling in the North Atlantic was concentrated mainly in a northern belt, across the ocean from the *Charleston* (number 5), *Hatteras* (number 4) and *Southern* (number 3) Grounds, through the *Western* (number 1), *Western Islands/Azores* (number 2) and *Steen* (number 8) Grounds, then along the eastern side of the ocean southwards virtually to the equator, through *San Antonio/Cape Verdes* (number 12) to *Cornell* (number 13). In the South Atlantic, there was a concentration of activity along much of the eastern South American coast, south through *Coast of Brazil* (number 18), *Platte* (number 19) to *False Banks* (number 20), and on towards the Falkland Islands (Unnamed Ground, number 21). There was also some activity in the central part of the ocean basin, particularly on *Pigeon* (number 24), *Tristan* (number 26) and to their northeast, on *Carroll/St Helena* (number 25).

There was considerable activity in the South Pacific. Whaling had moved up the South American west coast, as in the previous period, to major grounds off Peru (*Callao*, number 52), around *Galápagos* (number 38) and *Panama Bay* (number 35), west through the *Offshore* Ground (number 39), as well as west of *Panama Bay* (number 36), and along the equator through *On the Line (East)* (number 40) and further towards the west, through *On the Line (West)* (number 41). There was also considerable activity east of Australia in the Tasman Sea (*Middle*, number 45) and north of New Zealand (*Vasquez/French Rock*, number 46a), as well as towards *Fiji/Samoa/Tonga* (number 44). There was, however, a noticeable gap in operations in the open waters of the southeast Pacific.

Whaling in the North Pacific was largely centred on an east–west belt between the *Japanese Coast/Bonin Islands* area (number 31), through the *Japan* Ground (number 29) towards the *Hawaiian Islands* (number 32), but with little whaling between there and the equator. There was some activity in the East Indies, particularly in and around *Celebes Sea* and *Molucca Passage* (numbers 37b and 37c). Of two main centres of whaling activity in the Indian Ocean, one was in the west, off the African east coast, chiefly on the *Delagoa Bay/Mozambique Channel* (number 59), *East of Madagascar* (number 60) and *Mahe Banks* (number 58) Grounds and off Somaliland (Unnamed, number 56). The other was on the far side of the ocean, on the west and northwest coasts of Australia (mainly the *Coast of New Holland* Ground, number 62). Lesser centres included *Coast of Arabia* (number 55), *South of Ceylon* (number 61) and the central southern Indian Ocean around *Crozet Islands, St Paul/Amsterdam and Kerguelen Islands* (number 64).

Whaling in the Atlantic and Indian Oceans between 1850 and 1874 occurred in much the same areas as in the previous period. In the North Atlantic, there was rather more activity in the central ocean, on the *Twenty Forty* Ground (number 15). The same was true for the South Atlantic off western Southern Africa (*Coast of Africa*, number 22 and *Carroll*, including St Helena, number 25), as well as in the central ocean near Ascension Island (number 23). One exception was in the Indian Ocean where there are no extracts showing whaling off present-day Sri Lanka (*South of Ceylon*, number 61).

Whaling in the South Pacific also seems to have been prosecuted in much the same areas and at the same intensity as during the previous 25 years. But while there was rather more activity than earlier in the East Indies, particularly in the China Sea/*Sulu Sea* area (number 37a), the most noteworthy feature in the North Pacific is the marked decline in operations on the formerly active and extensive *Japan* Ground (number 29). Whaling in the North and South Atlantic between 1875 and 1922 seems to have continued at much the same level as previous years, and in much the same areas. Effort was rather more concentrated in the central North Atlantic, on the *Twelve-Forty* Ground (number 15), as well as in the South Atlantic on the *Coast of Africa* (number 22), both continuing the trend from the previous period. In the South Pacific there was a general decline in activity with only a small amount of whaling in scattered locations off western South America, north of New Zealand, and from the Galápagos Islands along the equator. The same applies in the North Pacific: there is a marked paucity of records overall, with the only concentrated activity in a small area on and north of the *Coast of Japan* (number 30). In the Indian Ocean, there were fewer records than earlier on the recognized East African and Western Australian grounds, but rather more activity in the southwest towards more southern latitudes.

Conclusions and further work

The picture derived from the Townsend extracts largely confirms the conclusions reached from the literature alone. Beginning in the North Atlantic during the early 1700s, sperm whaling expanded from the mid 18th century. The activity had moved from the Atlantic into the eastern South Pacific and southwest Indian Ocean just prior to the beginning of the 19th century. It peaked worldwide around 1840 to 1850, with activity in all oceans, and with major concentrations in the Pacific, especially in the North on the *Japan* Ground (number 29) – that is, between Japan and the Hawaiian Islands – and also along the Equator (*On the Line*, east and west, numbers 49 and 50). The second half of the 19th century was characterized by the abandonment of much of the North Pacific, particularly the formerly extensive *Japan* Ground, coupled with reduced operations in the western South Pacific and the western Indian Ocean. From around the 1880s, there was only scattered whaling in the Pacific and Indian Oceans. Yet, throughout that decline, rather remarkably, whaling seems to have continued throughout the Atlantic, even into the early part of the 20th century, especially in the South Atlantic on the *Coast of Africa* (number 22) and north of the equator on the *Twelve-Forty* Ground (number 15).

Although the general positions (and, in almost all cases, the names) of the various locations most favoured by the American sperm whalers did not change greatly over the century and a half of major 'open-boat' activity – the period circa 1750 to 1900 – their use certainly did. Economic and other industrial factors may well have played a part in those changes, but the assumption must be that they were driven largely by the availability of the whales. As Kugler wrote in 1971:

> *As vessels increased in number, the stocks of whales on the most accessible grounds declined. This was, of course, an ancient pattern ... in the 19th century, as in earlier times, the history of the industry can be seen as a history of the search for new and more productive whaling grounds, where the familiar types of whales could be exploited.*

There are, however, several kinds of possible error that warrant further examination. The discovery and sequential exploitation of areas where the whales aggregated at one season or another comprised the major theme of American sperm whaling. As noted above, Kugler saw this as driven by sequential depletion. That is consistent with observations on the *Japan* Ground (Bannister et al, 1981) and on the *Galápagos* Ground (Whitehead and Hope, 1991) of a fivefold decline in sightings and catches per day of whaling. On the other hand, Davis et al (1997) concluded that the industry's decline was more related to economic factors than to stock depletion. They did not include spatial aspects of the fishery's development in their analysis, however, and their results are more consistent with Whitehead's (2002) depiction of only a 30 per cent decline in abundance globally.

This critical appraisal of published (and some unpublished) information on the historical whaling grounds for sperm whales provides a basis for further analyses relating to the general question of causality in the decline of the fishery and, more broadly, to issues of population assessment. Overall, the extensive shifts in the whaling grounds used throughout the 19th century suggest that regional rather than global analyses are most appropriate and that the regional perspective might be rather different from the global. Specifically, following the methods used by Whitehead (2002) to scale up local or regional density estimates using Townsend's data to obtain a global estimate of current sperm whale abundance, it would be informative to characterize the physical and biological factors underlying the historical whaling grounds – for example, surface area; depths; bottom roughness or complexity; current patterns; and indices of primary and secondary productivity – more precisely. Similarly, it would be useful to examine if encounters with sperm whales were, in fact, more frequent on the grounds than in areas between them. Such analyses would give a more rigorous basis for the types of extrapolation made by Whitehead (2002).

It could also be useful to delve more deeply into the Townsend data to evaluate the typicality of the Townsend charts. Do the charts accurately reflect the entire American fishery, or are they biased towards certain ports or vessel riggings? Do they proportionally reflect catches over the entire two centuries of American open-boat sperm whaling, from the 1720s to the 1920s? There is also the question of 'spreading' in areas of especially dense plottings. Such spreading may be more extensive than Townsend himself claimed or than has been obvious previously, as suggested by the dark grey (Townsend chart positions) and light grey (actual positions from the Townsend extracts) dots in Figure 7.4 (overleaf). It should also be possible to refine the current understanding of degrees of 'depletion' of sperm whales on different grounds (Bannister et al, 1981; Whitehead and Hope, 1991) using the data considered here, supplemented by data extracted directly from whaling logbooks.

ACKNOWLEDGEMENTS

This chapter is a condensed version of an unpublished manuscript prepared by the lead author – Bannister, J. L. (2004) 'A study of published information on pre-20th century whaling grounds for sperm whales' – with the support of the Northeast Fisheries Science Center, Woods Hole, Massachusetts. Most generous facilities were provided to Bannister at the Kendall Institute, New Bedford Whaling Museum, by Stuart Frank and his staff, particularly Michael Dyer and Laura Pereira. The detailed analyses of the Townsend charts would not have been possible without the contribution of the Wildlife Conservation Society, especially Gillian Woolmer and Eric W. Sanderson. Anne Nevin, Margaret Triffitt and Wendy Crawford assisted Bannister at the Western Australian Museum, Perth, where facilities continue to be provided to him through the courtesy of the Trustees, the Executive Director and the Head of Science and Culture.

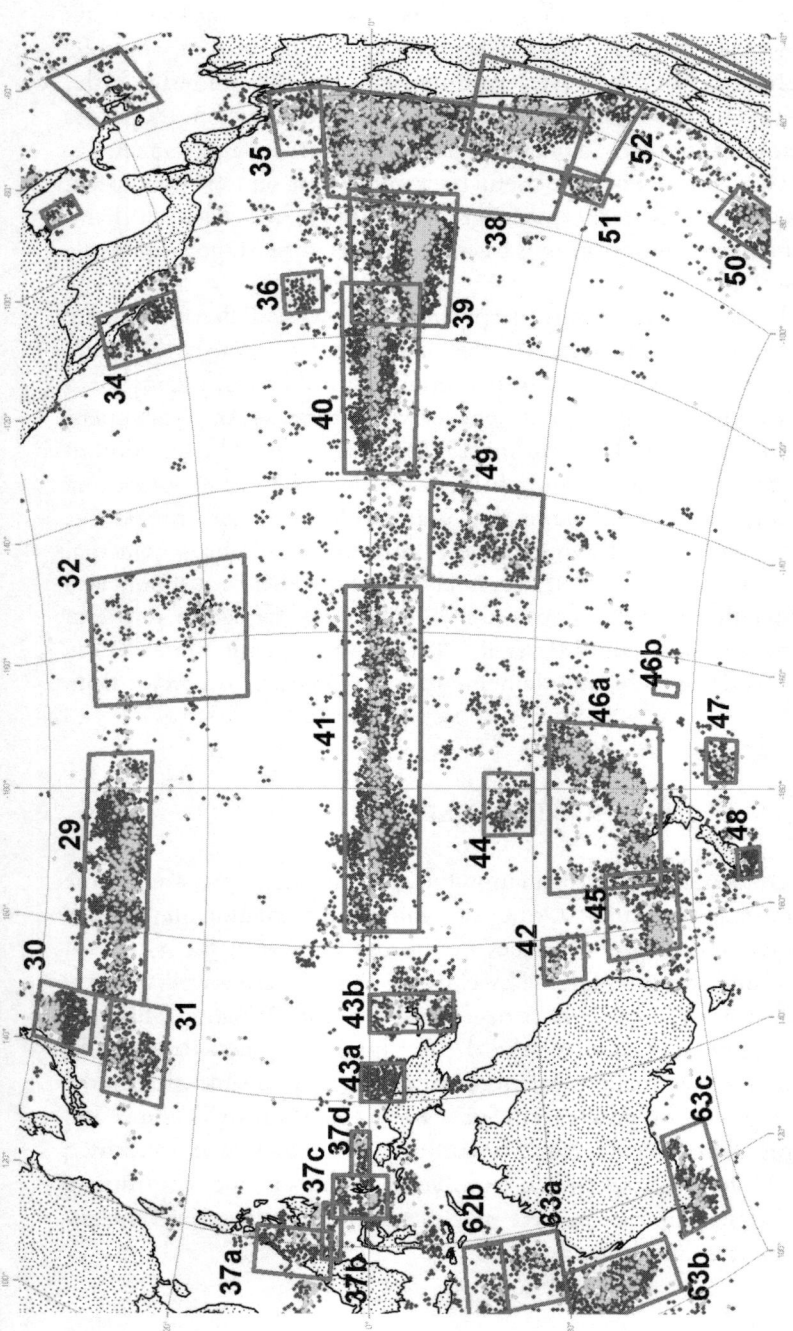

Figure 7.4 *An example of Townsend's sperm whale plots (Pacific Ocean)*

Note: Grounds numbered as in Table 7.3.

Source: digitized from Townsend's published charts (dark grey dots) and extracted from original logbooks (light grey dots)

References

Bannister, J. L. (1986) 'Notes on nineteenth century catches of southern right whales (*Eubalaena australis*) off the southern coasts of Western Australia', *Report of the International Whaling Commission*, Special Issue, vol 10, pp255–259

Bannister, J. L. and Mitchell, E. D. (1980) 'North Pacific sperm whale stock identity: Distributional evidence from Maury and Townsend charts', *Report of the International Whaling Commission*, Special Issue, vol 2, pp219–230

Bannister, J. L., Taylor, S. and Sutherland, H. (1981) 'Logbook records of 19th century American sperm whaling: A report on the 12-month project, 1978–79', *Report of the International Whaling Commission*, vol 31, pp821–833

Beale, T. (1835) *A Few Observations on the Natural History of the Sperm Whale*, Effingham Wilson, London, available online at http://mysite.du.edu/~ttyler/ploughboy/bealeold.htm, accessed 5 September 2007

Beale, T. (1839) *The Natural History of the Sperm Whale ... to which Is Added, a Sketch of a South-Sea Whaling Voyage*, John Van Voorst, London, available online at http://mysite.du.edu/~ttyler/ploughboy/bealenew.htm, accessed 5 September 2007

Broeze, F. J. A. (1977) 'Whaling in the Southern Oceans: The Dutch quest for southern whaling in the nineteenth century', *Economische-en Sociaal-Historisch Jaarbook*, Martins Nijhoff, Gravenhage, pp66–112

Cawthorn, M. W. (1995) 'Progress report on a re-evaluation of the Townsend historical whale catch data, with reference to bathymetry and other features', Meeting Document SC/47/5H29, Scientific Committee, International Whaling Commission, Cambridge, UK

Clark, A. H. (1887) 'The whale fishery. 1. History and present condition of the fishery', in *The Fisheries and Fishery Industries of the United States, Section V: History and Methods of the Fisheries*, Government Printing Office, Washington, DC, pp1–218

Davis, L. E., Gallman, R. E. and Gleiter, K. (1997) *In Pursuit of Leviathan. Technology, Institutions, Productivity and Profits in American Whaling, 1816–1906*, University of Chicago Press, Chicago

Ellis, R. (2002) 'Traditional whaling', in W. F. Perrin, B. Würsig and J. Thewissen (eds) *Encyclopedia of Marine Mammals*, Academic Press, San Diego, pp1316–1328

Gilmore, R. M. (1959) 'On the mass strandings of sperm whales', *Pacific Naturalist*, vol 1, pp9–16

Kugler, R. C. (1971) 'The penetration of the Pacific by American whalemen in the 19th century', *The Opening of the Pacific: Image and Reality*, National Maritime Museum, Maritime Monographs and Reports, no 2, pp20–27

Maury, M. F. (1851) *Whale Chart* (Preliminary sketch), Series F, National Observatory, Washington, DC, Sheet 1

Maury, M. F. (1852) *Whale Chart of the World* (The wind and current charts), Series F, Washington, DC, 4 sheets, no 1 (North Atlantic, Northeast Pacific) 1852; no 2 (Northwest Pacific) undated; no 3 (South Atlantic, Southeast Pacific) undated; no 4 (Southwest Pacific, Indian Ocean) undated

Reeves, R. R., Josephson, E. and Smith, T. D. (2004) 'Putative historical occurrence of North Atlantic right whales in mid-latitude offshore waters: "Maury's Smear" is likely apocryphal', *Marine Ecology Progress Series*, vol 282, pp295–305

Richards, R. (1982) *Whaling and Sealing at the Chatham Islands*, Roebuck Society, Publication no 21, Canberra, pp1–90

Richards, R. (2000) *Honolulu: Centre of Trans-Pacific Trade. Shipping Arrivals and Departures 1820 to 1840*, Pacific Manuscripts Bureau, Canberra and Hawaiian Historical Society, Honolulu

Scammon, C. M. (1874) *The Marine Mammals of the North-Western Coast of North America, Described and Illustrated, together with an Account of the American Whale Fishery*, J. H. Carmany, San Francisco, CA

Smith, T. D. and Reeves, R. R. (eds) (2003) *Design of a Program of Research on Sperm Whale Catch History: Results of a Workshop*, www.cmrh.dk/Cachalot.pdf, accessed 24 August 2007

Starbuck, A. (1964) *History of the American Whale Fishery from Its Earliest Inception to the Year 1876*, 2 vols, Argosy-Antiquarian Ltd, New York, NY

Townsend, C. H. (1931) 'Where the nineteenth century whaler made his catch', *Bulletin of the New York Zoological Society*, vol 34, no 6, pp173–179

Townsend, C. H. (1935) 'The distribution of certain whales as shown by logbook records of American whaleships', *Zoologica*, vol 19, no 1, pp1–50 + charts

Whitehead, H. (2002) 'Estimates of the current global population size and historical trajectory for sperm whales', *Marine Ecology Progress Series*, vol 242, pp295–304

Whitehead, H. and Hope, P. L. (1991) 'Sperm whalers off the Galápagos Islands and the western North Pacific, 1830–1850: Ideal free whalers?', *Ethnology and Sociology*, vol 12, pp147–161

Wilkes, C. (1845) 'Currents and whaling: The sperm whale fishery', *Narrative of the United States Exploring Expedition*, vol 5, pp487–493

Wray, P. and Martin, K. R. (1983) 'Historical whaling records from the western Indian Ocean', *Report of the International Whaling Commission*, Special Issue, vol 5, pp213–255

8

Depletion within a Decade: The American 19th-Century North Pacific Right Whale Fishery

Elizabeth A. Josephson, Tim D. Smith and Randall R. Reeves

The North Pacific right whale fishery was a remarkably short and bloody chapter in the history of whaling (Webb, 1988; Scarff, 2001). Right whaling had been conducted for centuries in the North Atlantic, and was practised in the South Pacific and South Atlantic, as well as the North Pacific, during the 19th century. The whale ships engaged were similar to those used for sperm whaling during the 19th century, and in the years immediately before and during the North Pacific right whale fishery, many of the same American vessels were also right whaling in the southwestern Pacific, as is reflected in the arrival and departure data for the Bay of Islands, New Zealand (Richards, 1992).

Commencing in the early 1840s, the North Pacific right whale fishery expanded rapidly. In the early 1850s, Lieutenant Matthew Fontaine Maury (1852 *et seq*) used data abstracted from American whaling logbooks to chart the worldwide distributions of sperm whales (*Physeter macrocephalus*) and right whales (*Eubalaena* spp). The heyday of right whaling in the North Pacific was still very recent, and Maury had access to abundant source material in the then-common voyage logbooks. These documents included detailed daily records that allowed him to plot the distribution and relative abundance of North Pacific right whales (*E. japonica*), which are now rare and endangered (Brownell et al, 2001; Shelden et al, 2005).

Maury's charts were based on two variables: the number of days on which whales were encountered, and the total number of days when whalers were present in an area. For two main reasons, Maury's simple index of relative abundance – whale encounters per day – needs to be interpreted with caution. First, by aggregating data from a range of years, Maury would probably have obscured marked changes in abundance that occurred as the fishery progressed. Due to such changes, his charts may under-represent pre-whaling abundance in areas that were searched more intensively in later years of the fishery, by which time whale numbers (and presumably density) had been reduced (Scarff, 1991). Second, errors appear to have been made in the preparation of the charts (Reeves et al, 2004), although these do not seem to have stemmed from flawed extraction of data from the logbooks. Fortunately, both the published charts and the unpublished summary sheets prepared by Maury's logbook readers, upon which the charts were based (the so-called *Maury Abstracts*), have been preserved and are available for examination.

This chapter attempts to address the first of these concerns by using the information that Maury and his assistants transcribed from the whaling logbooks to disaggregate the charted data by year, and by examining spatial shifts and fluctuations in whaling intensity as the fishery developed. Our analysis shows that in the course of a single decade – the 1840s – the once-numerous right whales in the North Pacific were rapidly and severely reduced in abundance, particularly in the eastern portion of their range.

Materials and Methods

The National Oceanic and Atmospheric Administration (NOAA), under the Comprehensive Ocean–Atmosphere Dataset (COADS) project, digitized selected data from all of the known surviving *Maury Abstracts*, including the daily vessel positions and thrice-daily weather observations of more than 525 multiyear whaling voyages (NOAA, 1998). Using the digitized positional data, we identified 270 voyages that included whaling in the North Pacific (north of 20° N). Whaling was seasonal, and a single voyage could involve two or sometimes three discrete time-limited periods of right whaling on the North Pacific grounds – here called 'vessel seasons'. The 270 voyages used for the present study involved some 465 vessel seasons, in which approximately 65,600 daily positions were given.

The *Maury Abstracts* constitute a sample of the total voyages undertaken to the North Pacific. The *Whalemen's Shipping List and Merchants' Transcript* (*WSL*, 1843–1914), a New Bedford whaling trade journal, published statistical data on the American fleet. We used the *WSL* to determine the full size and yearly pattern of the American whaling fleet in the North Pacific, and this allowed us to assess the extent to which Maury's sample was representative. We also used the annual oil returns from the North Pacific fishery published in the *WSL* to examine fluctuations in average yield per vessel.

We added to the North Pacific sections of the digitized daily positional data the information on whale encounters recorded in the *Maury Abstracts* (not included in the COADS project) for 187 of the 270 North Pacific voyages. The 83 voyages that we did not address were primarily those for which the abstracts were either illegible or failed to identify the whales to species. The 187 abstracts included a total of 9802 encounters with whales and 49,470 daily ship positions. The quality of the information in the abstracts on whale encounters is highly variable. Maury relied on several logbook readers who were inconsistent in the amount of detail that they recorded. The best abstracts contain information on the number of right whales seen, taken, and struck and lost. Unfortunately, many others simply record 'whale' or 'whales', with no indication of the species or whether the animal was taken or struck.

Using the positional information on encounters with right whales, we identified four regions of the North Pacific with clusters of sightings: the Gulf of Alaska, the east coast of Kamchatka, the area around the Kurile Islands and the Sea of Japan (see Figure 8.1). Right whale sightings north of 60° N were ignored for this analysis because of the problem of confounding right whales and bowhead whales (*Balaena mysticetus*) (Bockstoce and Botkin, 1983). The distinction between the two species was not firmly established until the 1860s and it is likely that some, many or most whales in the northern Bering Sea identified as right whales were, in fact, bowheads. We also included 'offshore' waters as a fifth region in our analysis, although relatively few whaling days, and encounters with right whales, were evident there.

We calculated encounter rates by dividing the numbers of days on which right whales were encountered by the numbers of days whaling vessels reported themselves to be in the region – that is, our encounter rates are simple proportions of days on which right whales were encountered. We computed confidence intervals based on the binomial distribution. To minimize biases due to inconsistencies in

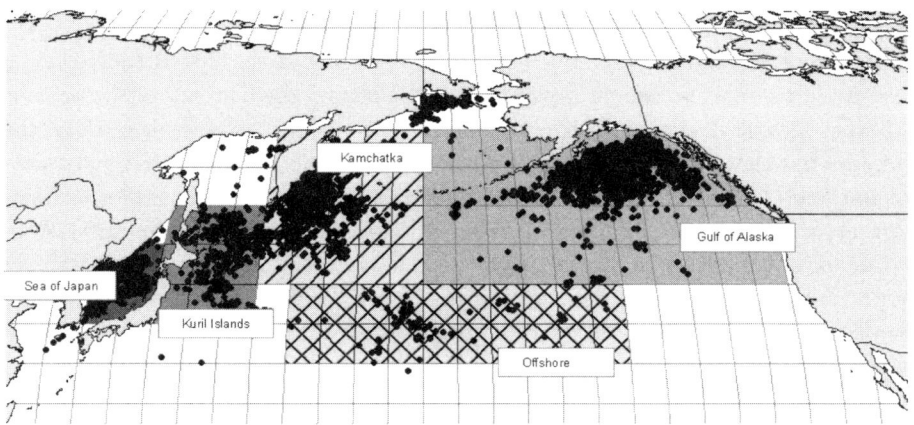

Figure 8.1 *Right whale encounters in the* Maury Abstracts *(dots) and regions selected for analysis*

Source: see text; view colour version at www.hull.ac.uk/oceanspast

the propensity of Maury's transcribers to record the species of whale encountered, we used the 96 abstracts in which more than 75 per cent of the whale observations were identified to species. These encounter data are from multiple vessels, some of which were operating in close proximity to each other. Some 'over-reporting' is therefore likely in that the same whales might have been reported in different vessels' logbooks as being sighted at approximately the same time and place. However, because both encounters and days on the whaling grounds in all 96 vessel seasons were included in calculating the rates, they should provide an unbiased measure of the likelihood of any vessel's encountering whales on a particular day.

RESULTS

Figure 8.2 shows the daily positions of whaling ships as recorded in the *Maury Abstracts* from 1821 to 1852. Although the 'Northwest Ground' in the Gulf of Alaska was discovered in 1835 (Kugler, 1986), large-scale commercial hunting of right whales in the North Pacific by offshore whaling fleets did not start in earnest until the 1840s. By 1844, significant numbers of vessels were reported as being 'on Kamchatka'; by 1845, whalers were venturing into the Sea of Okhotsk; by 1848, some were entering the Sea of Japan. However, when bowheads were discovered north of the Bering Strait in 1848, the American whaling fleet rapidly shifted to this new ground. Maury's sample clearly shows this pattern. It also shows the seasonal nature of this fishery, with whalers using the Hawaiian Islands as a staging area for the northern whaling operations.

Figure 8.3 shows the number of US vessels whaling in the North Pacific during each year as reported by the *WSL* from 1839 to 1853 (solid line), and the number of logbooks from those vessel seasons sampled by Maury (dashed line). The sampling fraction is relatively constant for most of the 1840s (dotted line) and it is clear that the fishery expanded rapidly during the first half of the decade. Maury's sample covers approximately 20 per cent of the fishery for most of the decade, but that percentage drops off after 1850. Whereas the scarcity of points in the track line plots during the 1850s reflects a decrease in the size of Maury's sample, the sharp increase in the number of vessel seasons during the early 1850s is due to the expansion of the Arctic bowhead fishery. The annual average catch per vessel in the North Pacific fishery in terms of barrels of whale oil, most of which came from right whales, is shown in Figure 8.4. The initial decline and subsequent increase in the 1840s was followed by a decline after 1850, probably reflecting spatial shifts in whaling activity from the eastern to the western Pacific and the ensuing development of the bowhead fishery (see Figure 8.2).

These spatial shifts broadly correlate with declines in encounter rate (percentage of days with right whale sightings) over time in each region (see Figure 8.5). In the Gulf of Alaska, although there was some initial fluctuation, rates that were greater than 50 per cent during the early 1840s had begun declining steadily by

DEPLETION WITHIN A DECADE 137

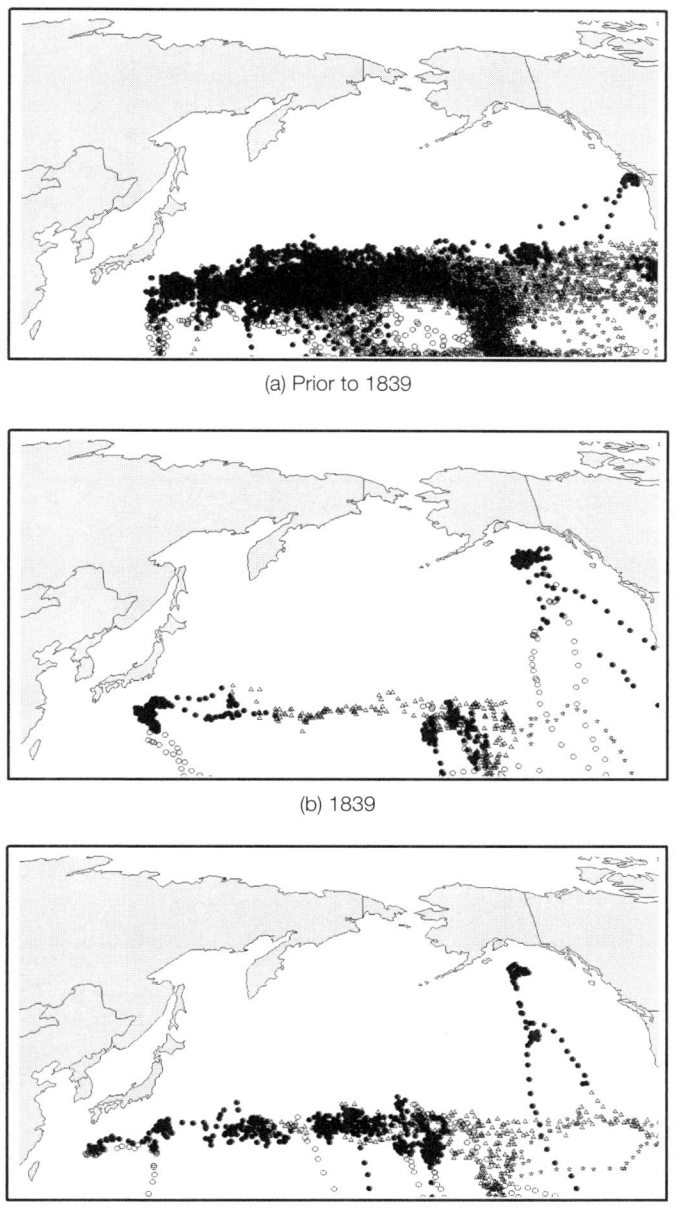

(a) Prior to 1839

(b) 1839

(c) 1840

Figure 8.2 (a–o) *Daily positions of whaling vessels in the* Maury Abstracts, *pre-1839 to 1852*

Source: derived from NOAA (1998); view colour version at www.hull.ac.uk/oceanspast

Notes: Daily positions are depicted as follows: ● = June–August; △ = September–November; ☆ = December–February; O = March–May.

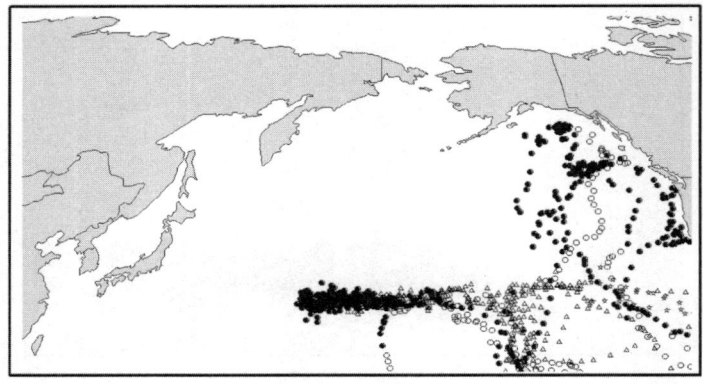

(d) 1841

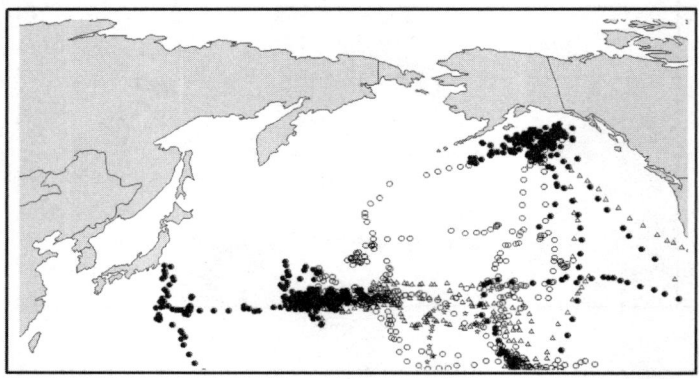

(e) 1842

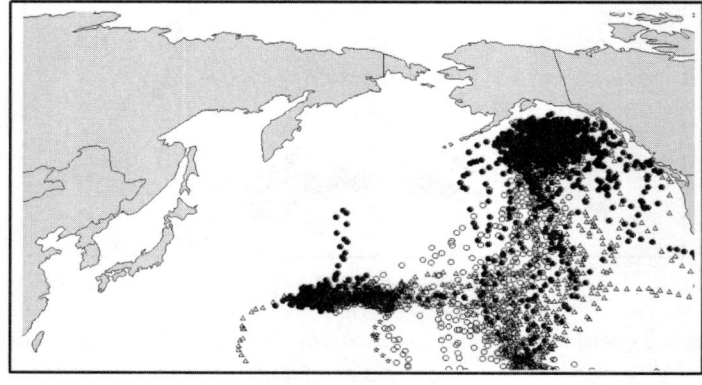

(f) 1843

Figure 8.2 *Continued*

DEPLETION WITHIN A DECADE 139

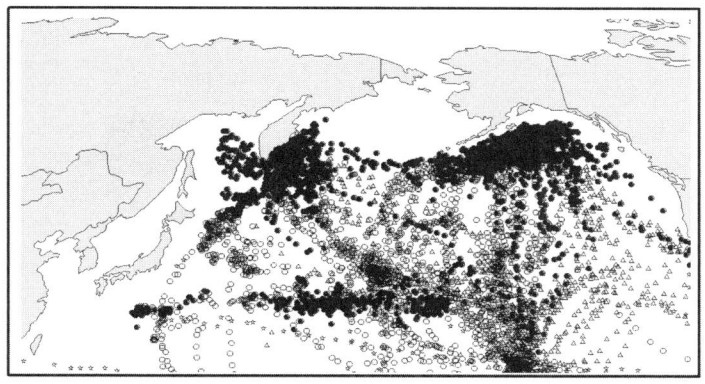

(g) 1844

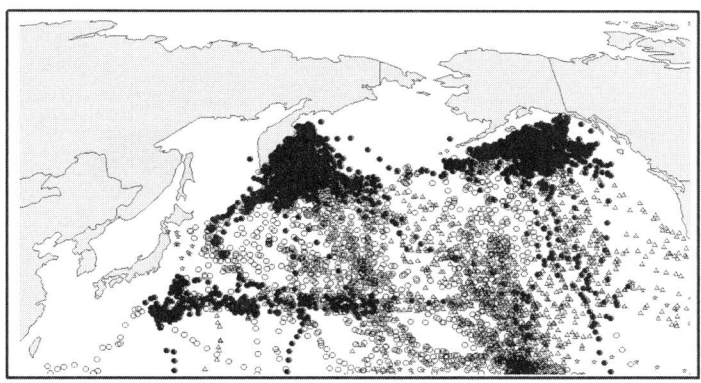

(h) 1845

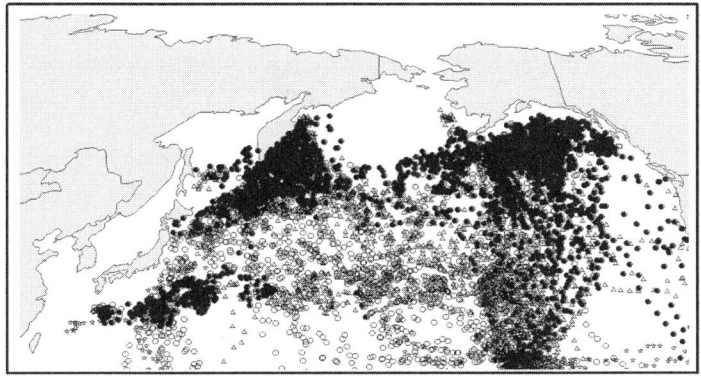

(i) 1846

Figure 8.2 *Continued*

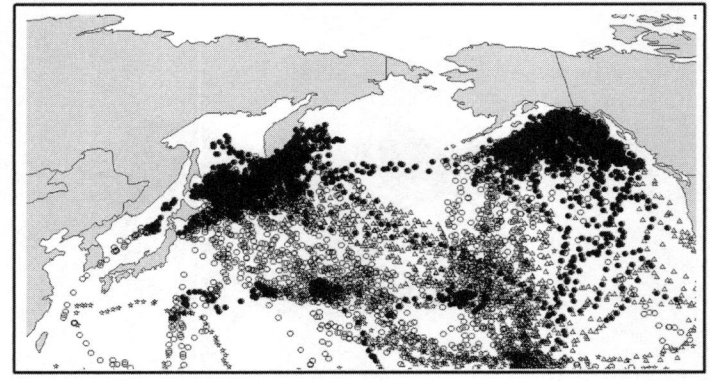

(j) 1847

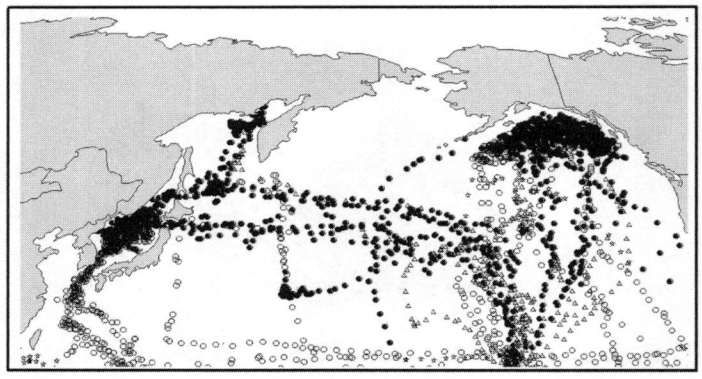

(k) 1848

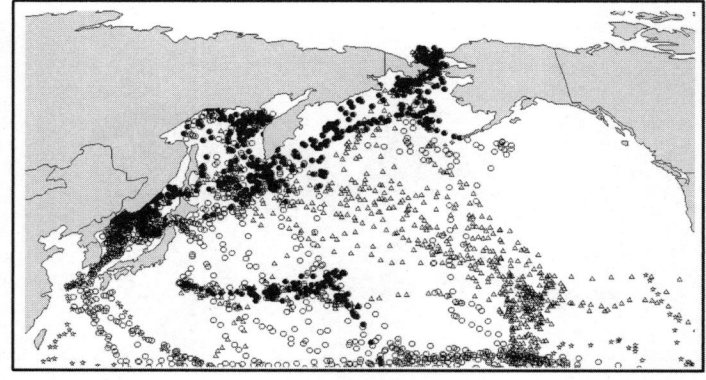

(l) 1849

Figure 8.2 *Continued*

DEPLETION WITHIN A DECADE 141

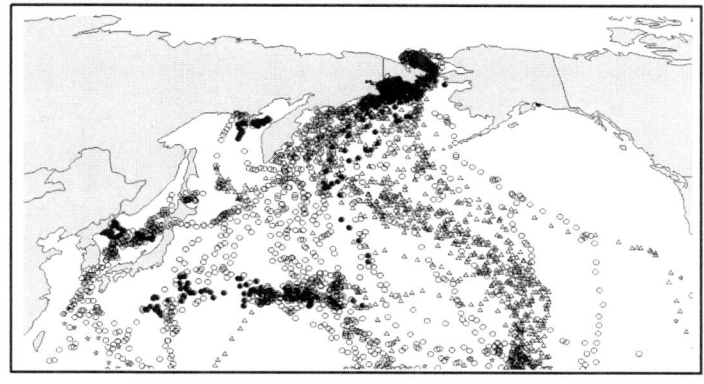

(m) 1850

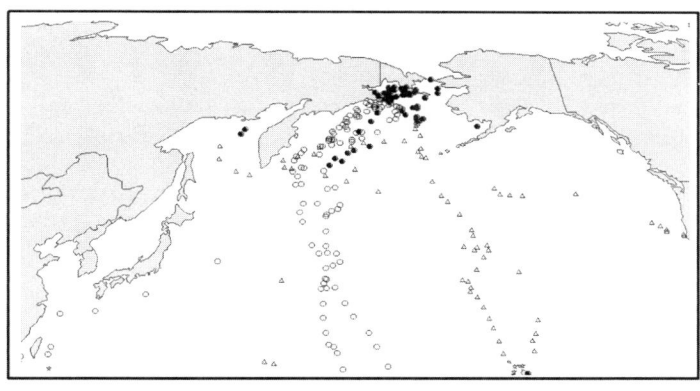

(n) 1851

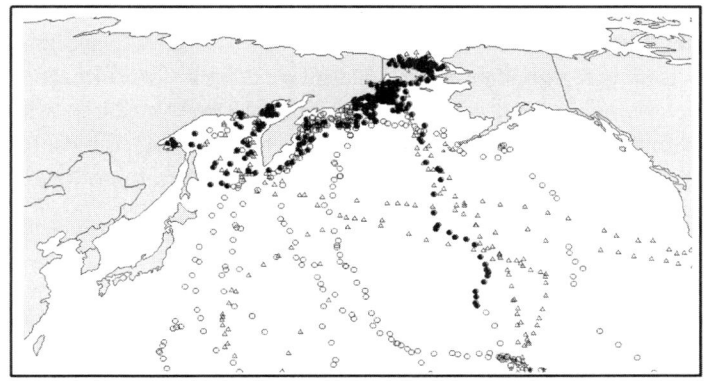

(o) 1852

Figure 8.2 *Continued*

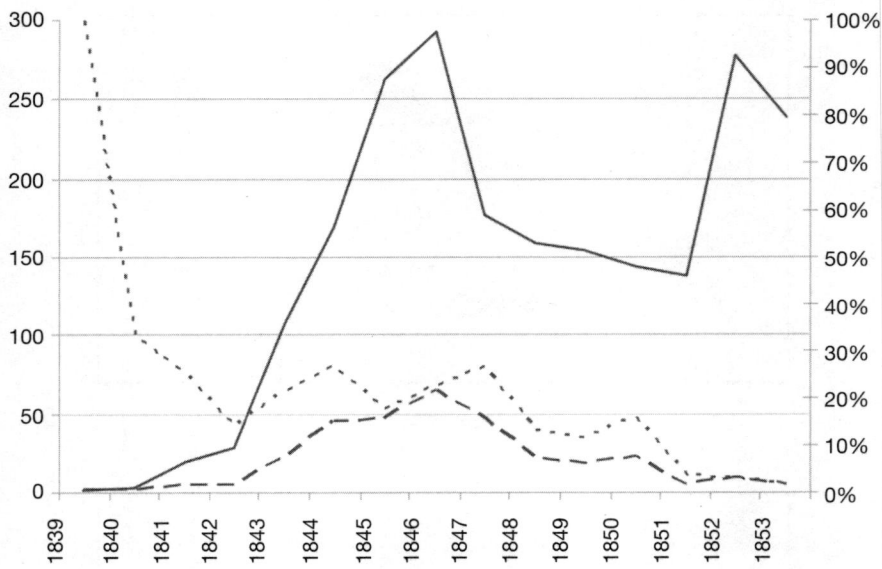

Figure 8.3 *Numbers of US vessels whaling in the North Pacific, 1839–1853*

Notes: Solid line, left axis: number of vessels reported by the *Whalemen's Shipping List*. Dashed line, left axis: number of logs of those vessels sampled by Maury. Dotted line, right axis: sampling fraction.

Source: see text

the middle of that decade, becoming only a small fraction of their initial levels by 1849. The coast of Kamchatka and the region around the Kurile Islands were being visited by 1844. Kamchatka encounter rates were initially greater than 40 per cent, but decreased rapidly thereafter and did not exceed 20 per cent after 1847. The Kurile Islands region experienced a similar pattern of decline, and the Sea of Japan, which was subjected to intensive whaling somewhat later, also exhibited quickly diminishing rates of encounter. Right whales appear to have been sighted only sporadically in the offshore zone, primarily in the spring months, and at rates considerably lower than those evident in the other areas.

Conclusion

The North Pacific right whale fishery was short lived. During the 1840s, a large American fleet of whale ships shifted into this region from sperm and right whaling elsewhere, and then soon moved on to the bowhead whaling grounds farther north. The overall picture is one of rapid escalation in the intensity of exploitation (see

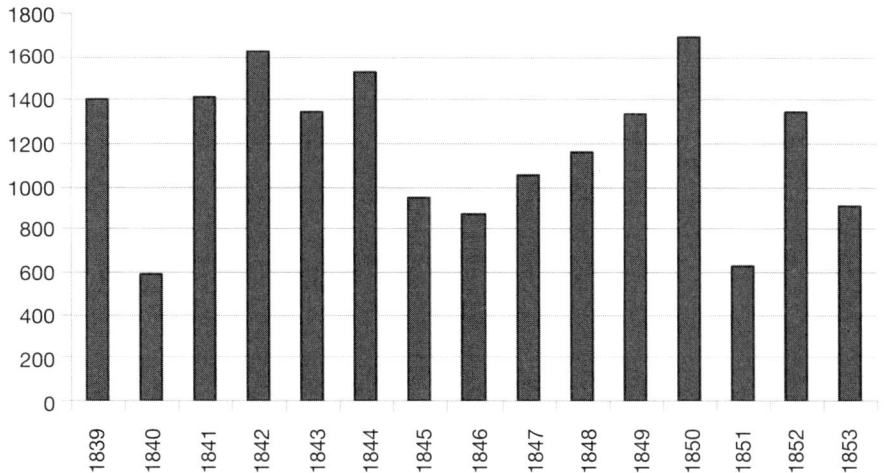

Figure 8.4 *Average catch reported by the* Whalemen's Shipping List and Merchants' Transcript *for the North Pacific fishery (barrels of oil)*

Source: Whalemen's Shipping List and Merchants' Transcript (1843–1914)

Figure 8.3), followed within just a few years by a rapid decline in encounter rates in the Gulf of Alaska, off Kamchatka, in the Kuriles and in the Sea of Japan (Figure 8.5). Some levelling off of encounter rates is evident in the western North Pacific at the end of the decade. In the Gulf of Alaska, rates declined after a secondary increase in effort.

The voyage tracklines imply a sequential opening, development and decline of fairly discrete whaling grounds, and the declines in whaling effort (Figure 8.3) correlate well with decreases in encounter rates on the various grounds. The encounter rates in the Gulf of Alaska ground began to decline in 1847, but the fleet began to expand spatially before that decline occurred. Apparently, none of the grounds could sustain removals at the rates that they were experiencing, and encounter rates eventually dropped in all of them.

From the voyage tracklines, it is clear that the Gulf of Alaska was the first area of the North Pacific where right whales were heavily exploited by the American fleet. After 1843, some whalers shifted from there to the western side of the North Pacific, off Kamchatka. Whaling pressure increased rapidly in the North Pacific fishery as the 29 American vessels operating in 1842 were succeeded by 108 in 1843 and 263 in 1845 (see Figure 8.3). The decline in average catches after 1844 and the fall in encounter rates in the Gulf of Alaska occurred as the effort intensified in Kamchatka. The expansion to Kamchatka and the Kuriles, and on to the Sea of Japan, did not sustain the industry for long – encounter rates soon dropped there,

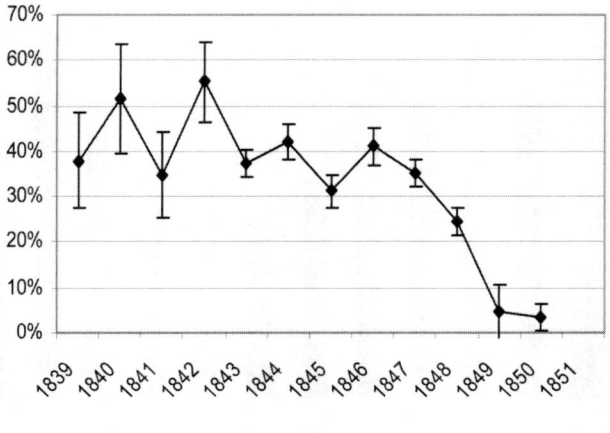

(a) Gulf of Alaska region

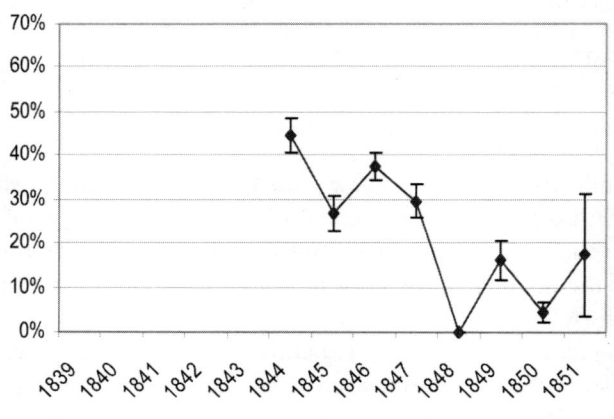

(b) Kamchatka region

Figure 8.5 *Right whale encounter rates 1839–1851, with 95 per cent confidence intervals*

Notes: (d) summer (June to August) encounter rates = black squares; spring (March to May) encounter rates = grey diamonds. (e) autumn (September to November) encounter rates = black squares; spring (March to May) encounter rates = grey diamonds.

Source: see text

DEPLETION WITHIN A DECADE 145

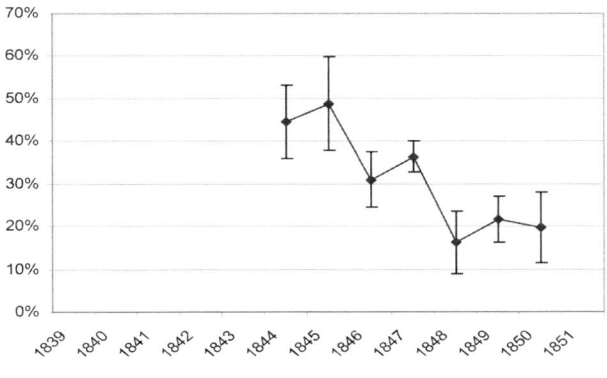

(c) Kurile Island region

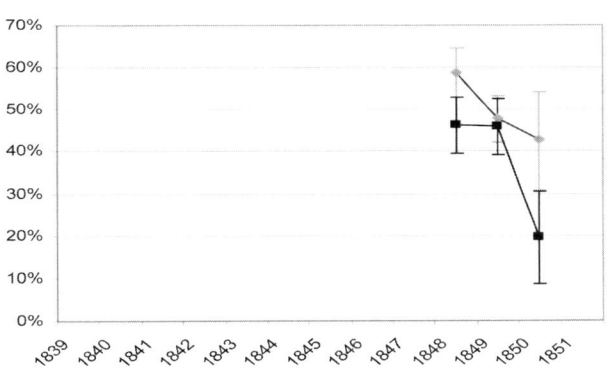

(d) Sea of Japan region

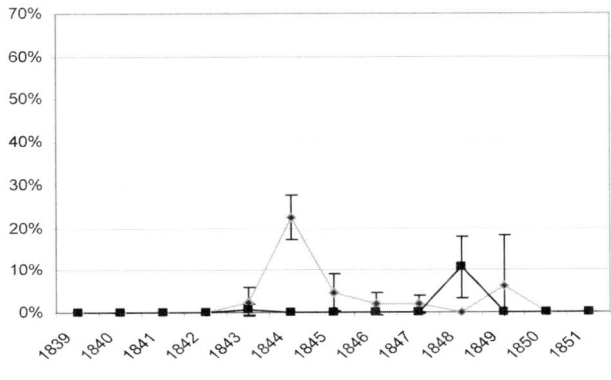

(e) Offshore region

Figure 8.5 *Continued*

as well. Although right whales were encountered farther south than the four main whaling grounds depicted by the aggregations shown in Figure 8.1, the intensity of whaling and the encounter rates in the offshore region – which were higher in the spring months than in other seasons – were substantially lower than in other areas (Figure 8.5e).

Because Maury's study was completed during the early 1850s, his data provide no information on subsequent developments, although it is well known that eventually right whales were nearly exterminated in the North Pacific (Brownell et al, 2001; Shelden et al, 2005). Encounter rates in the Gulf of Alaska had declined by an order of magnitude within ten years of the opening of the fishery, and over the next 50 years or so, the American fishery, although focusing primarily on bowheads, probably took a large fraction of the remaining right whales in the Gulf of Alaska and southeastern Bering Sea (Bockstoce, 1986, pp97, 151, 221). In the western North Pacific, right whale encounter rates had only declined by about half by 1849 according to the data analysed here. Further study of logbooks will be needed to enhance understanding of the further decline of right whales in the western North Pacific (including the Sea of Okhotsk) after 1850.

Our analysis of the *Maury Abstracts* provides a graphic, quantitative example of what has been described as a typical pattern:

> ... it is a parable of the whale fishery, now well understood, that each new whaling area yields up the bulk of its harvest in the first few years of whaling. Increased use only depletes the natural resources and leads to a failure of the ground. (Webb, 1988, p83)

As observed by many others, notably Scarff (2001), the boom-and-bust character of commercial whaling was compressed into a single decade in the case of the US-dominated North Pacific right whale fishery.

Acknowledgements

The work for this chapter was conducted under the auspices of the History of Marine Animal Populations programme, funded largely by the Alfred P. Sloan Foundation. Support was also provided by the Northeast Fisheries Science Center and the Office of Ocean Exploration, both of the US National Oceanographic and Atmospheric Administration. The authors are indebted to the US National Weather Service for assistance in accessing the *Maury Abstracts*, the New Bedford Whaling Museum for assistance in accessing whaling logbooks, and the Blunt-White Library of the Mystic Seaport for assistance in accessing the *Whalemen's Shipping List*.

REFERENCES

Bannister, J. and Mitchell, E. (1980) 'North Pacific sperm whale stock identity: Distributional evidence from Maury and Townsend charts', *Report of the International Whaling Commission*, Special Issue, vol 2, pp219–230

Bockstoce, J. R. (1986) *Whales, Ice, and Men: The History of Whaling in the Western Arctic*, University of Washington Press, Seattle, WA

Bockstoce, J. R. and Botkin, D. B. (1983) 'The historical status and reduction in the Western Arctic bowhead whale (*Balaena mysticetus*) population by the pelagic whaling industry, 1848–1914', *Report of the International Whaling Commission*, Special Issue, vol 5, pp107–141

Brownell, R. L. Jr., Clapham, P. J., Miyashita, T. and Kasuya, T. (2001), 'Conservation status of North Pacific right whales', *Journal of Cetacean Research and Management*, Special Issue, vol 2, pp269–286

Kugler, R. C. (1986) 'Report of the workshop on the status of right whales, Appendix 4: Random notes on the history of right whaling on the Northwest Coast', *Report of the International Whaling Commission*, Special Issue, vol 10, pp17–19

Maury, M. F. (1852 *et seq*) *Whale Chart of the World* (The Wind and Current Charts), Series F, Sheet 1 (North Atlantic, Northeast Pacific) 1852 and Sheet 2 (Northwest Pacific) no date, Washington, DC

NOAA (National Oceanic and Atmospheric Administration) (1998) *The Maury Collection: Global Ship Observations 1792–1910*, NOAA, National Climate Data Center, Asheville, North Carolina, US

Reeves, R. R., Josephson, E. and Smith, T. D. (2004) 'Putative historical occurrence of North Atlantic right whales in mid-latitude offshore waters: "Maury's Smear" is likely apocryphal', *Marine Ecology Progress Series*, vol 282, pp295–305

Richards, R. (1992) *Arrivals and Departures from the Bay of Islands, New Zealand, 1820 to 1840*, Paramata Press, Wellington, New Zealand

Scarff, J. E. (1991) 'Historic distribution and abundance of the right whale (*Eubalaena glacialis*) in the North Pacific, Bering Sea, Sea of Okhotsk and Sea of Japan from the Maury whale charts', *Report of the International Whaling Commission*, Special Issue, vol 41, pp467–489

Scarff, J. E. (2001) 'Preliminary estimates of whaling-induced mortality in the 19th century North Pacific right whale (*Eubalaena japonicus*) fishery, adjusting for struck-but-lost whales and non-American whaling', *Journal of Cetacean Research and Management*, Special Issue 2, pp261–268

Shelden, K. E., Moore, S. E., Waite, J. M., Wade, P. R. and Rugh, D. J. (2005) 'Historic and current habitat use by North Pacific right whales *Eubalaena japonica* in the Bering Sea and Gulf of Alaska', *Mammal Review*, vol 35, no 2, pp129–155

Webb, R. L. (1988) *On the Northwest: Commercial Whaling in the Pacific Northwest, 1790–1967*, University of British Columbia Press, Vancouver, BC

Whalemen's Shipping List and Merchants' Transcript (1843–1914) vols 1–72, 17 March 1843–29 December 1914, New Bedford, MA

9

Sperm Whale Catches and Encounter Rates during the 19th and 20th Centuries: An Apparent Paradox

Tim D. Smith, Randall R. Reeves, Elizabeth A. Josephson, Judith N. Lund and Hal Whitehead

On a global scale, the number of sperm whales caught during the 20th century greatly exceeded the 19th-century catch. However, in at least two whaling grounds – the Japan Ground in the central North Pacific and the Galápagos Ground in the eastern tropical Pacific – the rate at which American whalers encountered sperm whales during the 19th century declined rapidly over the years, suggesting significantly decreased whale abundance (Bannister et al, 1981; Hope and Whitehead, 1991). The discrepancy can be seen in Figure 9.1, which shows catches peaking and encounter rates declining markedly in the 19th century, with catches peaking again at much higher levels in the 20th century. Declines in encounter rates during the 19th century seem inconsistent with the fact that much larger catches were made on these and in adjacent regions during the 20th century (Tillman and Breiwick, 1983; Whitehead, 2002).

Two published analyses have used 19th- and 20th-century whaling data from the Pacific to compare population models, one for the Japan Ground (Tillman and Breiwick, 1983) and the other for both that ground and the Galápagos Ground (Whitehead, 1995), while a third assessment has examined whaling and population data at the global scale (Whitehead, 2002). All three assessments identified the same discrepancy between catches and encounter rates, but none was able to resolve it. Tillman and Breiwick's (1983) analysis indicated that the number of sperm whales in the North Pacific during the mid 20th century was substantially larger than it had been before significant whaling began in the 19th century, a conclusion that does not accord with conventional understanding of the effects of whaling.

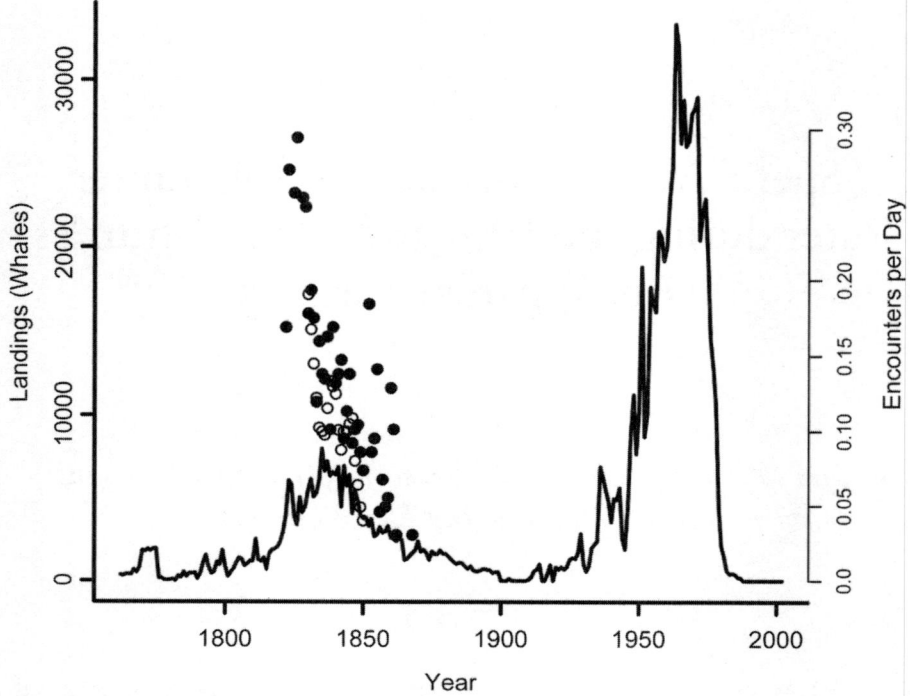

Figure 9.1 *Estimated annual global removals of sperm whales, circa 1750–2000, and encounter rates in the mid-19th century*

Note: Encounter rates from 19th-century whaling are denoted by closed circles on the Japan Ground and open circles on the Galápagos Ground.

Source: Japan Ground encounter rates from Bannister et al (1981); Galápagos Ground encounter rates from Hope and Whitehead (1991); 19th-century catches from Best (2005); 20th-century catches from International Whaling Commission; view colour version at www.hull.ac.uk/oceanspast

Examining both the Japan and Galápagos grounds, Whitehead (1995) enquired: 'If open-boat whaling so reduced sperm whale numbers, why were modern whalers able to take more whales over a shorter period (Best, 1983) from smaller initial stocks and not drive the populations to extinction?' Revisiting the problem at a global rather than regional scale, Whitehead (2002) concluded that 19th-century whaling had been intensive enough to inflict a moderate decline in the abundance of sperm whales without being sufficiently aggressive to reduce numbers to the extent suggested by the declines in 19th-century encounter rates. Thus, there remains a discrepancy, an apparent paradox, and it arises regardless of whether the catch data are considered regionally in the Pacific or at the global scale.

This chapter investigates this apparent paradox by exploring the implications of possible violations of four assumptions that underpin the three published analyses (see Table 9.1). First and most obviously, all three assessments, despite identifying uncertainties in the removal statistics, assumed that the estimated or reported catch levels were approximately correct. However, 20th-century removals may have been substantially lower, or pre-1900 removals substantially higher, than has been estimated. Examining the former possibility, Allison and Smith (2004) reviewed reported 20th-century catches. For the latter, information pertaining to 19th-century American whaling was assessed to determine whether the oil landing data used to estimate total catches and removals could have been much higher, or the yield of oil per whale considerably lower, than has generally been assumed.

The second assumption made by the analyses is that sperm whale populations were not only closed, but also affected by 19th-century whaling in the same

Table 9.1 *Four assumptions underlying the apparent paradox revealed by three analyses of 19th- and 20th-century sperm whaling*

Assumption	Ways Assumption Might Be Violated	Relevant Information
1. Removal estimates correct	20th-century catches over-reported	20th-century catch data from IWC
	19th-century oil landings underestimated	19th-century oil landings
	19th-century barrels per whale overestimated	19th-century American whaler logbooks
	Number of 19th-century voyages under-reported	Lists of 19th century American voyages
	19th-century whaling killed many more whales than were secured and processed	19th-century American whaler logbooks
	Paradox regional rather than global	Catches by ocean region
2. Separate regional populations or one global population	19th-century whaling of same populations on multiple grounds	Geographic location of 19th-century whaling grounds
	20th-century whaling of populations not subjected to 19th-century whaling	Geographic locations of 19th-century versus 20th-century whaling
	Whales changed their behaviour	Daily American voyage logbook data
3. Encounter rates proportional to whale abundance	Whalers changed their behaviour	Daily American voyage logbook data
	Heterogeneity in whale behaviour	Modern movement data from Galápagos Islands, and model
4. Equilibrium environmental conditions	Population carrying capacity changed	Alternate model formulations
	Ocean climate changed	Climate time series

Source: Tillman and Breiwick (1983); Whitehead (1995, 2002)

way – that is, there were no refuges. Both Tillman and Breiwick (1983) and Whitehead (1995) assumed that during the 19th century, sperm whales formed isolated populations that were only subjected to catching effort on single grounds. Whitehead (2002), in contrast, treated all sperm whales as though they belonged to a single global population. To explore the possibility of multiple populations, catch data by ocean region for both centuries were analysed and distances between the various 19th-century whaling grounds were considered in the light of known migration and distribution patterns. The spatial overlap between catches during the 19th and 20th centuries was then evaluated to determine whether refuges from 19th-century whaling activity might have existed.

A third assumption made by Tillman and Breiwick (1983) and Whitehead (1995) was that whale encounter rates derived from logbook data (as in Bannister et al, 1981, and Hope and Whitehead, 1991) are reliable indicators of actual declines in whale abundance. However, the quantitative interpretation of catch and encounter rates is difficult for several reasons. For example, just as whalers, like all fishermen, might alter their searching strategy over time, their prey could have responded to whaling pressure by changing their avoidance and flight behaviour, as well as their distribution. Indeed, Whitehead (2002) suggested that 'the discrepancy between the severe drop in the sighting rates of the whalers over this period [1830–1850] and the results of the model may be at least partially explained if the whales changed their schooling behavior or distribution as exploitation... progressed'. In addressing this issue, data from the Japan Ground were analysed to compare encounter rates and interpret changes over time and space, while evidence from both the Japan and Galápagos grounds was assessed to judge whether the proportion of encountered whales that were killed changed over time. Another potential cause of mismatch between whale abundance and encounter rate is behavioural heterogeneity. If some whales were easier to find and kill because of their behaviour, the rate of encountering individuals would have declined more quickly than the population as a whole, which over time would include a higher proportion of the more elusive animals. Sperm whale populations in the South Pacific are culturally structured into 'clans' (Rendell and Whitehead, 2003) that consistently behave in different ways (Whitehead and Rendell, 2004). Whether such differences could explain the apparent paradox was investigated using modern data and theories regarding the movements of sperm whales of different clans.

Fourth, according to all three analyses, environmental constancy is assumed to have prevailed during the 1800 to 2000 period. Tillman and Breiwick (1983, p267) acknowledged that a significant increase in carrying capacity of 'the North Pacific sperm whale population' was a 'possible speculation', but found 'no information giving this supposition any great substance'. Whitehead (1995) also considered this possibility; but in his 2002 paper he did not raise the issue, implicitly assuming a constant environmental carrying capacity for sperm whales globally. At least two types of change in the environment could have occurred. On the one hand, ecological relationships may have altered as whale numbers were reduced by whaling – for example, the competitive balance between sperm whales and their unexploited competitors for squid could have led to disequilibrium conditions (cf. May et al, 1979). On the other hand, climate change may have affected the inherent productivity of the oceans over the two centuries. Of course, these two types of change may have interacted as well.

In the sections that follow, these four assumptions, and several alternative hypotheses, are examined in detail. The results of several specific comparisons and suggestions for further research are then discussed in the concluding part of the chapter.

Removal estimates

Two distinctly different sources of data on sperm whale catches are available for the periods before and after circa 1900. During the 18th and 19th centuries, the catches were recorded in terms of the amount of sperm whale oil produced by the fishery, while most 20th-century catches were assembled by the Secretariat of the International Whaling Commission (IWC) and reported as numbers of individual animals killed and processed. A painstaking review of data sources by the secretariat has not revealed any substantial problems with the 20th-century data, although several instances of misreported and under-reported catches have come to light in recent years. Allison and Smith (2004) concluded that, at least in the case of sperm whales, such problems were insignificant in comparison to the overall magnitude of the catches. The reported catches for most whaling operations were judged reliable; in any event, the more likely downward bias would be in the wrong direction to help resolve the apparent paradox.

Estimates of sperm whales killed prior to 1900 are much more complicated and the uncertainties greater. Best (2005) developed estimates of total catches of sperm whales by dividing the volume of oil imported by the yield of oil obtained from an average whale. For American whaling returns, Best relied on tables of landings published in Starbuck (1878), Hegarty (1959) and a major trade newspaper, the *Whalemen's Shipping List and Merchants' Transcript* (*WSL*). To evaluate whether Best could have underestimated 19th-century American catches to such an extent as to explain the apparent paradox, two new datasets were constructed. First, drawing on Lund (2001), a new list of American whaling voyages, which included the volume of sperm oil produced by each voyage, was assembled. The annual total volume of oil landed from all voyages during the year that they were completed was then compared to the total yearly imports of sperm oil into the US. Second, data on the yield of oil per whale were compiled from logbooks and compared to the average oil yield values used by Best (2005).

Oil production based on American whaling voyages

In this analysis, voyages are defined according to customs house records of departure and arrival because that was the basis for reported landings. We supplemented Lund's (2001) comprehensive and meticulously documented list of US whaling voyages with data relating to landings and whaling grounds derived from voyage lists (Wood, 1831–1873), inbound customs manifests (especially for New London, Connecticut) and logbooks and journals (Sherman et al, 1986), as well as information in Starbuck (1878), Hegarty (1959) and Davis et al (1997). Starbuck (1878) did not describe his methods or sources; but in a flier circulated to potential collaborators (Sherman et al, 1986), he acknowledged his reliance on the New Bedford Shipping List. By this he presumably meant the *WSL*, a trade newspaper published from 1843 to 1914 that attempted to cover the entire industry, not just the fleet sailing from New Bedford. For vessels from New Bedford, the most active port, we relied on production data assembled by Davis et al (1997).

Lund (2001) had identified some voyages unknown to Starbuck (1878) and Hegarty (1959), and we further checked the voyage data by comparing the information given in more than one source (see Table 9.2). Where possible, inconsistencies were resolved and errors corrected, while data on some voyages that had not appeared in Lund (2001) were added. The final database included nearly 15,000 voyages that sailed from 112 US ports from the early 18th century to circa 1900.

The annual totals of the amount of sperm oil produced by each voyage were compared to the yearly quantities of oil imported that Best (2005) had derived largely from Starbuck (1878, Table I) and Hegarty (1959). It is evident that Starbuck (1878, Table I), and presumably also Hegarty (1959), had, in turn, reproduced the *WSL* annual totals of sperm oil landed by port from 1843 to 1914, the only differences being minor transcription errors. For this period, there were substantial year-to-year differences between the *WSL*'s annual total imports and the corresponding totals of our voyage-based landings. One cause of divergence was that oil freighted during a voyage appeared in the import totals in the year that it was received, whereas in our voyage data, freighted oil was assigned to the year during which the voyage was completed. Indeed, for this entire period, the annual differences nearly averaged out (see Table 9.3), with the total oil imported being slightly less (1.8 per cent) than the total oil estimated from individual voyages. Similarly, Starbuck (1878, Table I) gave annual total oil importation by port for the 38 years (1804 to 1842) prior to the publication of the *WSL*, although he did not indicate his sources. Again there were annual differences between those import aggregates and our voyage-based total production, with the former 4.5 per cent greater than the latter. The differences in the measures of total oil production for pre- and post-1843 are not consistent with the proportion of voyages without reported production figures: 15.2 per cent and 8 per cent for the two periods, respectively.

Table 9.2 *Information regarding 18th- and 19th-century American open-boat whaling voyages derived from published sources*

Information	Lund (2001)	Starbuck (1878)	Davis et al (1997)	Sherman et al (1986) [1]
Captain name	•	•	•	•
Vessel name	•	•	•	•
Rig type	•	•	•	•
Home port	•	•	•	•
Tonnage		•	•	•
Day depart		•		•[2]
Month depart		•	•	•[2]
Year depart	•	•	•	•[2]
Day return[3]		•		•[2]
Month return[3]		•	•	•[2]
Year return[3]	•	•	•	•[2]
Geographic locations		•[4]	•[5]	•[6]
Product information		•	•	
Logbook existence	•[7]			•[7]

Notes: 1 Covers all voyage logbooks and journals in most public collections.
2 Dates covered in logbook or journal.
3 Dates of vessel's return to port or of loss of vessel.
4 One of more than 125 destinations recorded at the beginning of each voyage.
5 One or more of roughly 25 destinations derived from the destinations recorded at the beginning of each voyage.
6 One or more of 12 oceanic regions where the logbook indicated 'significant whaling'.
7 The present-day location of logbooks and journals pertinent to a voyage.

Table 9.3 *Aggregate sperm oil imports to US ports and voyage-based oil production*

	Pre-WSL Publication	During WSL Publication
Period	1803–1842	1843–1876
Aggregate oil imports (1000s of bbls)	2595	2645
Voyage-based oil production		
Barrels of oil (1000s of bbls)	2477	2692
Number of voyages represented	4110	6376
Percentage of all voyages	84.8	92.0

Source: aggregate oil imports from Starbuck (1878, Table I); voyage-based oil production data generated by methods described in text

Three factors may have contributed to the differences between the import-based and voyage-based total oil production in Table 9.3:

1 accuracy of the estimated figures for the barrels of oil;
2 oil production for voyages that were incompletely reported; and
3 completeness of the reporting of freighted oil.

To explore these factors, we compared our voyage information with additional voyage-specific information reported in the *WSL* (see Appendix 1 at the end of this chapter) and arrived at the following conclusions about each of the three factors. First, the amount of oil landed for each voyage was initially estimated at the end of each voyage and measured subsequently, barrel by barrel. The differences between estimated and measured landings for individual voyages in two years examined, 1849 and 1859, were small (less than 2 per cent) and not systematic. Second, the average production of voyages where the amounts of oil landed were missing was probably lower than for voyages where those data were available. This is indicated by the fact that 19 voyages ending in 1859 were reported by Starbuck but not by the *WSL*, and these landed only 31.7 per cent as much per voyage as the voyages that were reported in both sources. Third, while most of the 20.8 per cent of the sperm oil production shipped back to New England could be assigned to specific voyages, this was not true for 14 per cent of the oil from vessels that embarked in 1859. Such oil was unlikely to be represented in the voyage-based totals.

These findings suggest that 19th-century sperm whaling returned more oil than has generally been assumed, and that the difference could be approximated by adjusting the totals of the voyage-based oil production (see Table 9.3) to account for the voyages without recorded output (15.2 per cent for 1804 to 1842 and 8.0 per cent for 1843 to 1878). However, such adjustments would have to account for their likely lower production, which was perhaps as low as 31.7 per cent of that returned by other voyages. Furthermore, 20.8 per cent of all oil was freighted and as much as 14.7 per cent of such oil was probably not assignable to a voyage. If these proportions are representative, the voyage-based landings may be biased downward by as much as 7.7 per cent during 1804 to 1842 (= 15.2 per cent × 31.7 + 20.8 per cent × 14 per cent), and 5.5 per cent during 1843 to 1878 (= 8.0 per cent × 31.7 + 20.8 per cent × 14 per cent). However, the magnitude of these corrections might well have varied over time, and additional data for other years would be necessary to adjust the total figures more precisely.

Oil yield per whale

Best (2005) estimated the landed catch of whales by dividing the total oil imports as discussed above by the average amount of oil obtained from individual whales (yield). In order to obtain average yields for over 800 voyages, he divided the total barrels produced (Starbuck, 1878; Hegarty, 1959) by the number of whales landed (Townsend, 1935). For voyages that returned more than ten sperm whales, the average of this ratio was greater for whales taken by ships and barques (33.6 barrels, or bbls) than for those taken by brigs and schooners (19.1 bbls).

We explored this further using information in logbooks as recommended by Best (1983), selecting those cases in which the amount of oil obtained from one or more whales processed on board had been reported. This included data extracted for more than 60 complete voyages (under Josephson's supervision), the Japan Ground sections for 100 voyages (Bannister et al, 1981) and the Galápagos Ground entries for more than 30 voyages (Hope and Whitehead, 1991). The average yield was significantly lower ($p = 0$) for schooners than for vessels with other rigs (see Table 9.4), but the averages for all vessel types were larger than those obtained by Best (2005). For example, these new estimates, assuming that the sample sizes in Table 9.4 reflected the relative contribution of the different rigs, would imply an average yield that was 41.5 per cent higher. Because this enters the calculations as a divisor, the net effect would be to reduce estimated catches by 29.3 per cent. The cause of the differences between our estimates and Best's (1983, 2005) is not apparent, although Best identified a number of uncertainties in his data sources. Further examination of Townsend's (1935) logbook extractions, and the voyage production figures in Starbuck (1878) and Hegarty (1959), together with more logbook research, is needed to resolve the differences and to determine more precisely the variability of the yield by both vessel type and geographic region.

Hunting losses

Hunting loss was a feature of all whaling operations, and therefore the total number of whales actually 'removed' was undoubtedly larger than the catches alone. Nineteenth-century whalers sometimes struck (harpooned, lanced or shot) whales that they were unable to secure and process. Not all of the so-called 'struck and lost' component of the 'catch' died. This is known because some previously struck whales were subsequently killed and secured, their bodies bearing scars, wounds, lines and harpoons from encounters with whalers. However, some previously struck whales did die of their wounds, as demonstrated by logbook entries noting that some whales were found dead with harpoons still attached. The carcasses that were not too putrefied were processed for their oil, which was included in the voyage production figures. This would have partially offset the struck-and-lost mortality. Accordingly, 'correcting' the catch data (whether number of individuals processed

Table 9.4 *Average oil yield of sperm whales reported in 19th-century logbooks for voyages by vessels of different types*

Vessel type	Average (bbls)	Sample size (individuals)	Standard error of the average (bbls)
Bark	46.1	342	1.47
Brig	43.1	72	3.33
Schooner	31.3	200	1.55
Ship	45.9	946	0.74

Source: see text

or amounts of product obtained) to account for hunting loss is clearly neither simple nor straightforward.

Bannister et al (1981) concluded on the basis of 19th-century logbook data on the Japan Ground that landings should be increased by at least 20 per cent, but no more than 61 per cent, to account for hunting loss of sperm whales. Similarly, Hope and Whitehead (1991) used the same method to estimate hunting loss in the Galápagos Ground at between 4 and 23 per cent. They suggested that their lower loss rates might be explained by the fact that the sperm whales hunted in tropical waters were smaller and easier to catch. Indeed, based on our logbook data, the average yield from Galápagos Ground whales was independent of vessel rig ($p = 0.47$) and smaller (31.4 bbls, 1.10 standard error) than the average yield for all observations (see Table 9.4). The issue is more complex, however, because our logbook data for more than 60 complete voyages show that the proportion of the 7795 animals struck and lost differed significantly ($p = 0$) for whales encountered alone or with others (11.8 per cent, 0.39 standard error, and 18.3 per cent, 1.19 standard error, respectively). These observations suggest that the proportion of struck whales that were lost varied substantially by ground and by the nature of the group of whales encountered. Further investigation of the numbers reported struck and lost on different whaling grounds is required to clarify this issue.

Were 19th-century removals underestimated?

The foregoing analyses suggest that 19th-century sperm whaling landed more oil than has generally been assumed. While oil production has probably been underestimated, the average amount of oil per whale may also have been underestimated. This latter factor could more than offset the former. Furthermore, the appropriate factor for 'correcting' catch data to account for hunting loss appears to vary between whaling grounds and perhaps with the size of the whales being hunted. Despite the uncertainties, it is unlikely that the paradox can be explained solely by the underestimation of removals during the 19th century.

POPULATION STRUCTURE

The assessment by Tillman and Breiwick (1983) posited that the Japan Ground 'corresponds to a major portion of the presently defined Western Division of the North Pacific sperm whale population', as then recognized by the IWC. Similarly, the assessment by Whitehead (1995) assumed that whales taken on the Japan Ground and on the Galápagos Ground belonged to separate populations and that neither was subject to whaling on other grounds. Whitehead's (2002) global assessment made no attempt to incorporate population structure and thus implicitly assumed a single worldwide population of sperm whales, although he considered some alternatives in the form of sensitivity tests.

Table 9.5 *Estimated number of sperm whales removed (thousands)*

Region	1700s	1800s	1900s
North Atlantic	25	48	39
South Atlantic	0	23	14
Indian Ocean	<1	30	64
South Pacific	0	110	110
North Pacific	0	60	283
Antarctic	0	0	64
All regions	29	271	721

Source: see text

We explored an intermediate spatial scale between the regional populations of the first two assessments and the global population of the third using Best (2005) and Allison and Smith (2004) to allocate the global catch estimates by century into six large oceanic regions (see Table 9.5). Twentieth-century catches were less than or equal to 19th-century catches in three of the regions: the North Atlantic, South Atlantic and South Pacific. Twentieth-century catches in the Indian Ocean were only about twice those of the earlier era, while the catches in the North Pacific were more than 4.5 times larger, and those in the Antarctic were restricted to the 20th century. These differences suggest that the apparent paradox relates more to the Antarctic and North Pacific than to other areas. Referring to the Antarctic, Best (1979) argued that catches there were mainly of males that migrated seasonally to lower latitudes. The populations to which such whales belonged therefore would have been subjected to 19th-century whaling. With regard to the North Pacific, the relationship between encounter rates and catches was markedly different from that evident in the South Pacific (see Figure 9.2).

Twentieth-century whaling grounds in the North Pacific overlapped with, and extended north of, those used during the 19th century (see Figure 9.3). Because the oceanography of the North Pacific is more complex than that of the Southern Hemisphere (Best, 1979), it cannot be assumed, by analogy to the Antarctic, that sperm whales found in high latitudes of the North Pacific would have migrated seasonally to lower latitudes where sperm whaling occurred during the 19th century (mostly south of 40° N). Indeed, Kasuya and Miyashita (1988) inferred from 20th-century sightings, catches, movements of marked whales and oceanographic data that there are two populations of sperm whales in the North Pacific which occur mainly north of 40° N latitude, one in the west and the other in the east. In the eastern North Pacific, data from 20th-century shore whaling stations have been interpreted to indicate that sperm whales mated (April to May) and calved (July to August) off British Columbia (Gregr et al, 2000), well north of the 19th-century sperm whaling grounds. Accordingly, the region north of 40° N may have been a refuge from 19th-century whalers. It is nonetheless perplexing, given the

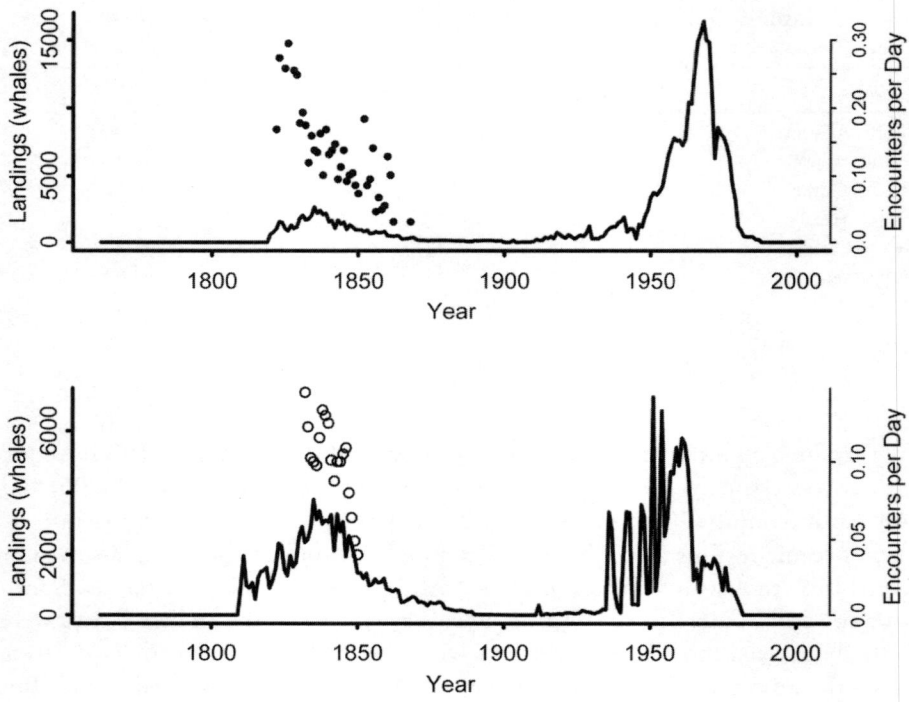

Figure 9.2 *Estimated annual removals of sperm whales, circa 1750–2000, for North Pacific (top) and South Pacific (bottom)*

Note: Encounter rates from 19th-century whaling are denoted by closed circles on the Japan Ground and open circles on the Galápagos Ground.

Source: see text

large numbers of sperm whales taken north of 40° N during the 20th century (see Figure 9.3), that so few sperm whales were reported from this region in whalers' logbooks. For example, in data compiled by Maury from 59 logbooks from the 1840s (see Chapter 8 of this volume), sperm whales constituted only 17 of 3791 sightings of whales identified to species during 59 voyages reporting in 10,995 days north of 40° N. In addition, given the potential of sperm whales to move over long distances (Best, 1979), many of the 19th-century Pacific whaling grounds would have been close enough to one another (see Figure 9.3) to allow some animals from the population using the Japan Ground or the Galápagos Ground to have been taken on other grounds. These removals would not have been included in the assessments by either Tillman and Breiwick (1983) or Whitehead (1995). To explore this possibility as a way of explaining the paradox, estimates of removals from the other whaling grounds are needed.

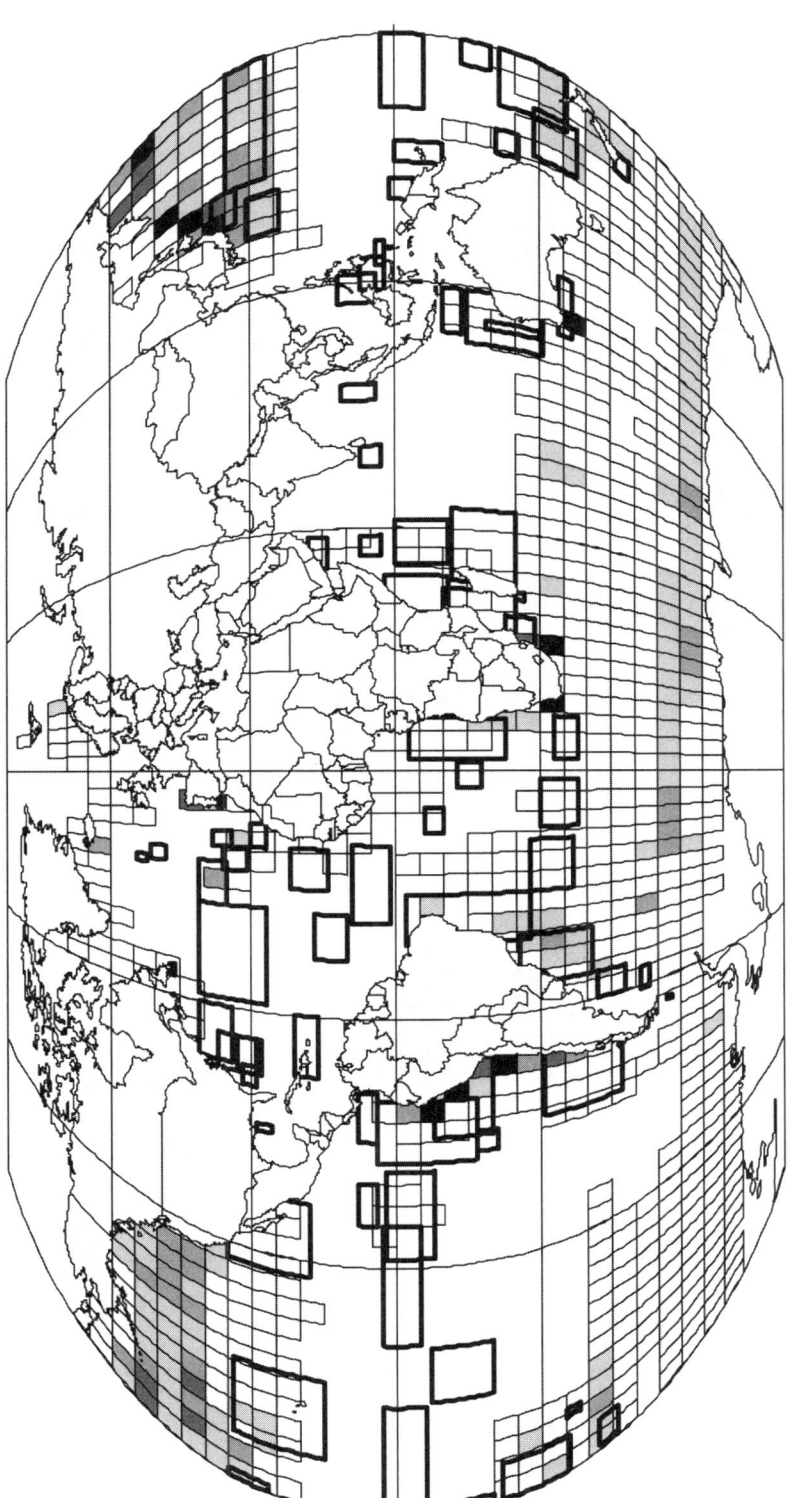

Figure 9.3 *The locations of 20th-century catches shown by shaded 5° squares of latitude and longitude*

Note: Darker shading depicts larger catches; 19th-century whaling grounds are shown by rectangles bounded with darker lines.

Source: see text; view colour version at www.hull.ac.uk/oceanspast

ENCOUNTER RATES PROPORTIONAL TO WHALE ABUNDANCE

The rates at which whalers encountered sperm whales in the Japan and Galápagos grounds declined as whaling continued (see Figure 9.2). Such declines were the norm according to many authors. For example, Herman Melville (1851, p458) wrote: 'In former years these Leviathans ... were encountered much oftener than at present.' The apparent paradox has arisen because declines in encounter rates have been interpreted to imply that whale numbers also declined. As mentioned earlier in this chapter, Whitehead (2002) suggested that the correlation between encounter rates and whale numbers might not be as close as is generally assumed, and that the paradox could be explained in part by changes in the whales' 'schooling behaviour or distribution as exploitation ... progressed'.

We made estimates of encounter rates on the Japan Ground using a sample of logbook data from the Maury Abstracts (Josephson et al, this volume). They exhibited a pattern similar to that found by Bannister et al (1981), with both datasets showing a possible initial increase followed by a long-term decline (see Figure 9.4).

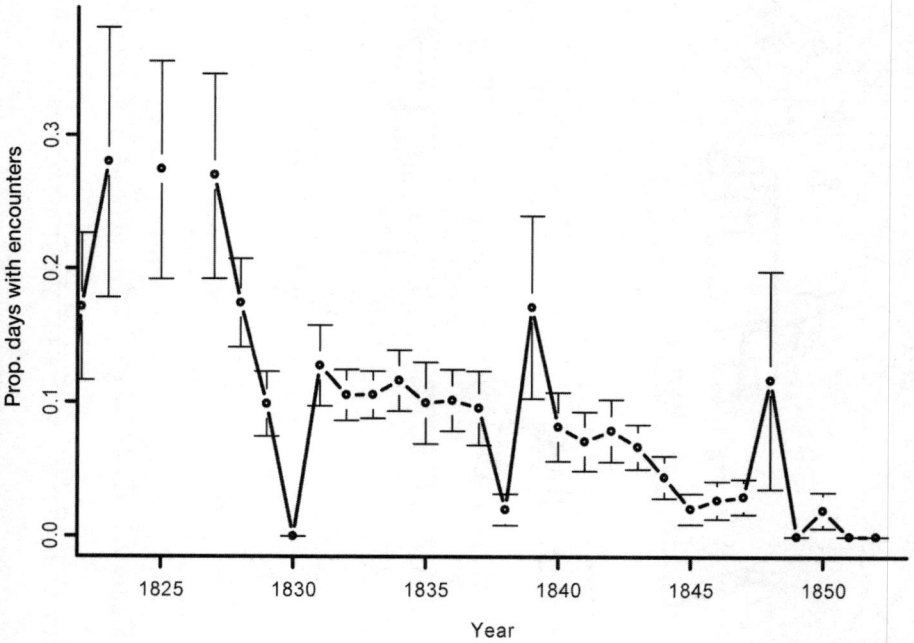

Figure 9.4 *The annual proportion of whaling days with encounters of sperm whales on the Japan Ground, 1822–1852*

Source: data derived from American whalers' logbooks, as extracted for Maury (1852 *et seq*), with 95 per cent confidence intervals

If sperm whales altered their behaviour after exposure to whaling, then the rate of encounter might have changed faster than abundance. This could have occurred if the animals became more evasive as they gained experience of whalers. However, this possibility is more about tendencies than absolutes as sperm whales continued to be taken, albeit at decreasing rates; moreover, not all individuals learned to avoid whaling ships. For example, the log of the whaler *Sarah* of Edgartown recorded that on 5 February 1850, at 27° S and 55° W: 'saw a skule [sic] of sperm whales ENE. Saw some of the whales past within 20 rods of the ship that would have made 100 bbls and some come within 10 feet of the ship. They were all lob tailing and breaching all around the ship'.

A partial test of the increasing tendency of the whales to avoid the whalers is to determine if the success in striking or securing animals once sighted decreased over the years. For the period of 1830 to 1850, we examined the ratio of whales taken to whales sighted for each voyage over time for both the Japan Ground (using data from Bannister et al, 1981) and the Galápagos Ground (using data from Hope and Whitehead, 1991). The ratio declined over time (–0.0076, $p = 0.02$) for the former ground, but did not decline (0.0051, $p = 0.61$) for the latter ground. These results are contradictory, but the nature of the two grounds was rather different and the fishery on the Japan Ground progressed westward over time. Moreover, even if the results for the two grounds had been consistent, they still would have represented only a partial test because the whales may have learned to avoid whale ships from a considerable distance before they were within range of being sighted.

The whales could also have changed their distribution. For example, Melville (1851, p458) suggested such changes when he argued against interpreting encounter rates as indicating a decline in abundance:

> *And equally fallacious seems the conceit, that because the so-called whale-bone [baleen] whales no longer haunt many grounds in former years abounding with them, hence that species also is declining. For they are only being driven from promontory to cape; and if one coast is no longer enlivened with their jets, then, be sure, some other and remoter strand has been very recently startled by the unfamiliar spectacle.*

Although Melville was referring to other species of whales, we examined our encounter rate data for evidence of a shift by sperm whales into areas adjacent to the Japan Ground. There was no increase in encounter rates over time (see Figure 9.5). However, this test is complicated because, as is apparent in Bannister et al (1981), the focus of whaling on the Japan Ground shifted westward over the course of the fishery. A more spatially resolved analysis is therefore needed to account for the effect of that shift.

Other changes in the searching and catching processes may have caused encounter rates to decline more slowly than abundance, or indeed even to increase over time regardless of any trend in abundance. Although such changes would

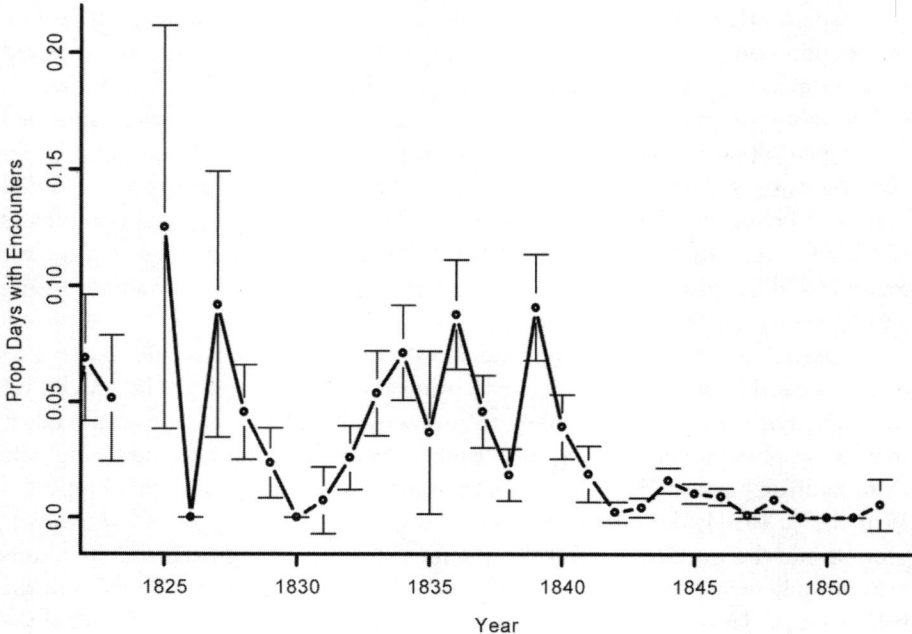

Figure 9.5 *The proportion of whaling days with encounters of sperm whales in areas adjacent to the Japan Ground, 1822–1852*

Source: data derived from American whalers' logbooks, as extracted for Maury (1852 et seq), with 95 per cent confidence intervals

not help to explain the apparent paradox, they might have tempered the effect of any alterations in the behaviour of whales. For example, searching effort might have become more efficient over time. The distribution of whales is not uniform spatially, with greater concentrations in some regions than in others. Differences in distribution can occur at small spatial scales relative to the distances whalers searched (at least several tens of miles per day). Whalers clearly used accumulated information about successful whaling grounds, as can be seen in lists and manuscript notebooks from the 19th century, such as are found in the Nicholson Collection of the Providence (Rhode Island) Public Library. Unpublished manuscripts with titles such as *Abstracts of the Latitude and Longitude of Whaling Grounds Right and Sperm 1871–1883* and *Whaling Grounds, 1840–1881* make it clear that whalers left port armed with information based on more experience than just their own (see Maury, 1854). Accordingly, they may have become better at finding whales as they gained experience on a particular ground. This is suggested, for example, by the initially increasing encounter rates in Figure 9.4, a feature that is also suggested in Bannister et al's (1981) independent sample of encounter rates for this same ground. Examining changes in encounter rates for the early years on several whaling grounds might reveal if this initial increase was a common pattern.

Another factor affecting encounter rates is the time spent processing killed whales. When encounter rates were higher, more whales were caught (Bannister et al, 1981), implying that proportionally more time would have been devoted to towing, cutting, boiling blubber, packing oil and other processing tasks. Logbooks do not distinguish consistently between hunting (searching, chasing and killing) and processing time, so the only measure of hunting time is days on the whaling ground. Because vessels tended not to move during at least the initial processing of a whale, it is conceivable that fewer whales would have been sighted on the days immediately following a catch. Accordingly, the number of whales encountered per day might have been lower when catches were higher and higher when catches were lower. This would have the effect of levelling out trends in encounter rates over time. Such effects might be detected by comparing the numbers of encounters per day in the days immediately before and after a catch.

Heterogeneity in whale behaviour also might lead to encounter rates declining faster than population size if one segment of the population is being encountered and killed preferentially. Off the Galápagos Islands, female sperm whales principally belong to one or the other of two clans with quite different movement patterns (Whitehead and Rendell, 2004). If whalers are moving at speeds comparable to, or slower than, those of the whales, their encounter rates will depend upon the movement patterns of the whales. Thus, for slow-moving open-boat whalers, the differential movement patterns of the clans might explain the sperm whale paradox. We used the actual tracks of groups of whales in each clan followed off the Galápagos Islands in 1987 and 1989 (Whitehead and Rendell, 2004) to calculate expected encounter rates by whalers moving at different speeds (see Appendix 2). The differences in encounter rates are real (about 15–20 per cent for whalers moving at one knot, and 5–7 per cent for whalers moving at two knots), but even these unrealistically slow speeds make only a tiny difference (a factor of about 1.0008 for whalers moving at two knots) to the reduction in sighting rate as compared to the contraction in true population size (see Appendix 2 at the end of this chapter). In fact, a huge difference in sighting rates between the two types of animals (about sixfold; see Appendix 2) would be needed to produce a decline in sighting rate twice as large as the decline in population size. It is therefore unlikely that behavioural heterogeneity resolves the sperm whale paradox.

Taken together, these possible changes in the behaviour of whales or whalers would have served to either increase or decrease the rate of decline of encounter rates. Further investigation is needed to determine if the net effect is in one direction or the other. In any event, there is no evidence to suggest that this factor alone would explain the paradox.

Equilibrium population dynamics

An underlying assumption of most analyses is that populations would tend to increase towards their pre-whaling levels – the environmentally determined carrying capacity – with the decline and eventual cessation of whaling. This concept is at best an approximation, and it presumes the reversibility of changes in the ecosystem due to decreases in whale population sizes, as well as a lack of long-term changes in oceanic productivity due, for example, to climate change. Although such an approximation may be reasonable over a short time frame, its validity over two centuries is less credible. For example, changes in population dynamics of humpback whales in the North Atlantic were addressed in a recent modelling study (Punt et al, 2007), where the assumption of constant carrying capacity was relaxed to determine if that would resolve apparent inconsistencies between the model's output and the empirical evidence for that population. Similar consideration of models that do not make equilibrium assumptions may be useful in the present case to determine if they would help to resolve or remove the apparent paradox.

Conclusion

The paradox appears to apply to sperm whales in the North Pacific. Large discrepancies in catches between the 19th and 20th centuries occurred there and in the Antarctic. Animals taken in the latter region almost certainly belonged to populations that were subjected to 19th-century whaling in lower latitudes, at least seasonally. It is possible, however, that in the North Pacific the whales subjected to 20th-century whaling to the north of 40° N latitude belonged to populations that were not affected significantly by whaling during the 19th century. This would mean that relatively unexploited refuge populations existed in the North Pacific during the early 20th century. If these were sufficiently large to have supplied enough whales to account for the very high 20th-century catches, the apparent paradox might be at least partially explained.

Some other possible explanations of the apparent paradox (see Table 9.1) can be essentially ruled out. For example, it is unlikely that 20th-century removals have been greatly overestimated. Further, it is unlikely that 19th-century removals have been underestimated to an extent that would explain the paradox, although the available estimates are subject to several forms of uncertainty that do not entirely preclude this. Some of the biases are upward and some are downward, but the magnitudes are generally small. One exception could be hunting loss, although further information on struck-and-lost rates on specific whaling grounds is needed. It also seems unlikely that behavioural heterogeneity within sperm whale populations would be the sole explanation for the paradox.

Further possible explanations warrant more research. One is the possibility that the behaviour of either the whales or the whalers, or possibly both, changed. The mixed results of the partial test of behavioural change do not allow this possibility to be ruled out. Additional data from logbooks may help with the interpretation of changes in encounter rates that are crucial elements of the apparent paradox. We also cannot rule out the possibility that the problem is rooted in unrecognized population structure or, alternatively, in the incorrect assumption of equilibrium conditions. Further studies are needed to improve definitions of sperm whale populations and provide better understanding of long-term environmental variability. New and emergent research tools that are being applied to these issues, such as genetic analyses and satellite-linked tagging and tracking of individual whales, will, in time, greatly improve the state of knowledge. In the meantime, attention should be devoted to developing estimates of 19th-century removals from additional Pacific whaling grounds and identifying long time series of data for use in assessing environmental change. These types of data could be used with population modelling to determine which, if any, combinations of population structure and environmental change explain the apparent paradox.

APPENDIX 1: COMPLETENESS OF VOYAGE-BASED DATA

We obtained additional data on individual voyages from the weekly issues of the *WSL* for 1849 and 1859 to examine the completeness of the voyage data assembled. First, relative to the accuracy of the landings data themselves, while most of Starbuck's voyage landings corresponded to entries in the *WSL* that were measured in port in preparation for sale (marked as 'gauged'), some values were preliminary estimates given by the captain when the vessel entered port. This was probably because the preliminary values were published in the *WSL* immediately upon arrival, and the gauged values became available only later. In order to determine the accuracy of the preliminary values, we extracted both preliminary and measured values for all voyages ending in 1849 and 1859. We compared these values pairwise and determined that there was no systematic difference and, further, that the average difference was less than 2 per cent in both years.

Second, undocumented and poorly documented voyages were probably ones with lower-than-average production. To investigate this question, we examined in detail the voyages reported by Starbuck and by the *WSL* as having ended in 1859, a sample year at the height of the sperm whale fishery (see Table 9.6). During that year, Starbuck reported that 194 voyages produced a total of 86,431 barrels of sperm oil (oil landed by the ship plus oil freighted or 'sent home'). We matched those voyages as closely as possible to the 176 voyages reported by the *WSL* for that year, which produced 83,357 bbls. The 18-voyage difference reflected 19 voyages that appeared only in Starbuck and one that appeared only in the *WSL*.

Completed voyages (i.e. voyages with a specified date for return to the port of departure) that were reported in both Starbuck and the *WSL* had very similar stated production. The freighted oil amounted to 20.8 per cent of the total production.

Table 9.6 *Sperm oil landed and freighted for voyages ending in 1859*

		Starbuck (1878, Table K)				Whalemen's Shipping List			
		Number of voyages	Oil landed (bbls)	Oil freighted (bbls)	Oil per voyage (bbls)	Number of voyages	Oil landed (bbls)	Oil freighted (bbls)	Oil per voyage (bbls)
Complete voyages	Both Sources	165	68,333	13,115	493.6	165	67,431	13,793	492.3
	One Source	4	89	45	33.5	1	0	0	0
Incomplete voyages	Both Sources	10	0	2133	213.3	10	0	2133	213.3
	One Source	15	0	2716	181.1	0			
Total		194	68,422	18,009	445.5	176	67,431	15,926	473.6

Source: Starbuck (1878, Table K); WSL

Table 9.7 *Sperm oil freighted and the number of shipments arriving in New England in 1859*

Port of origin	Vessel Type	Assigned to voyage			Not assigned to voyage		
		Number of shipments	Oil shipped (bbls)	Oil per shipment (bbls)	Number of shipments	Oil shipped (bbls)	Oil per shipment (bbls)
Known	Freighter	24	13,058	544.1	3	886	295.3
	Whaler	9	5149	572.1	0		
Unknown	Freighter	0			11	1482	134.7
	Whaler	16	2007	125.4	12	935	77.9
Total		49	20,214	412.5	26	3303	127.0

Source: see text

The four completed voyages reported in Starbuck but not in the *WSL* had much lower production, and the one voyage in the *WSL* but not in Starbuck did not return any sperm oil. The remaining voyages (25 reported by Starbuck and 10 by the *WSL*) were not completed back to the port of departure – the vessels were either lost, condemned or sold during the voyage. The incomplete voyages had no recorded landings, of course; but significant amounts of oil were freighted. For example, Starbuck's 25 incomplete voyages freighted an average of 194 bbls, substantially more than the average amount freighted by the 165 complete voyages (79.4 bbls). The average production of the 19 voyages reported only by Starbuck was 150 bbls, 31.7 per cent of the average production of all voyages reported in the *WSL*.

Third, to learn more about how freighting oil back to New England may have affected our voyage-based data, we identified in the *WSL* 75 shipments of whaling products (oil and baleen) arriving in 1859 aboard freighters and whaling vessels, and attempted to match the oil in these shipments with voyages in Starbuck's list. We were unable to assign 26 of these 75 shipments (34.7 per cent) to the voyages that produced the oil, and it is probable that Starbuck was likewise unable to attribute these shipments to specific voyages. They accounted for only 3303 of the 23,513 barrels (14.0 per cent) of oil shipped (Table 9.7). This difference arose because the average amount of oil per shipment differed according to whether the shipment could be assigned to a voyage. The averages also differed for shipments according to whether the port of origin was known or unknown. Specifically, the average was much smaller for shipments that could not be assigned to voyages than for those that could (127.0 versus 412.5 bbls).

APPENDIX 2: BEHAVIOURAL HETEROGENEITY AND ENCOUNTER RATES

Movement patterns and encounter rates

We calculated the proportional difference in the encounter rates of groups of sperm whales of the 'regular' and 'plus-one' clans (Rendell and Whitehead, 2003) off the Galápagos Islands in 1987 and 1989 for whalers moving at different speeds. We considered days spent following groups with positions at 06:00, 09:00, 12:00, 15:00 and 18:00 hours when the group being followed was either of the regular or plus-one clan. We linearly interpolated a track between these positions and drew a grid of resolution 0.5 nautical miles (nm) around the track. For each whaling vessel's speed (0, 1, 2, 3, 4, 5 knots) and direction of travel (in 15° increments relative to true north), we calculated the number of cells of the grid in which such a vessel could have been at 06:00 hours and seen the whales at some point during daylight (06:00 to 18:00 hours) given a range of visibility of three nm (results were robust to changes in this). The total area of these cells is proportional to the encounter rate for whalers heading in that direction at that speed. The areas (in square nautical miles) of the 06:00 positions of whalers who encountered whales are averaged over the headings for that day to give an overall encounter rate for each speed.

The results indicate that if the whaling vessels moved at an average of 2 knots, then the plus-one clan groups were visible at a rate about 5 to 7 per cent greater than the regular clan groups (see Table 9.8). This diminishes with increasing vessel speed and increases sharply for vessels moving more slowly.

Table 9.8 *Encounter rates of different clans of sperm whales for whaling vessels moving at different speeds*

Year	Clan	Days	Whaler speed (knots)					
			0.0	1.0	2.0	3.0	4.0	5.0
1987	Reg.	19	141.8	160.3	212.3	277.9	347.5	417.6
	Plus-1	7	172.0	184.8	223.3	284.4	351.8	420.8
	Ratio: Plus-1/Reg.		1.213	1.153	1.052	1.023	1.012	1.008
1989	Reg.	8	124.3	154.8	210.9	276.6	346.6	417.0
	Plus-1	5	170.9	185.3	224.9	284.7	351.9	420.8
	Ratio: Plus-1/Reg.		1.375	1.197	1.066	1.029	1.015	1.009

Source: see text

Rates of decline in encounter rates

If groups of one clan are encountered proportionally more often than groups of the other, how does this affect the decline in encounter rates with exploitation? Supposing there is initially a proportion p of clan 1 encountered at rate α_1, and $q = 1 - p$ of clan 2, encountered at rate α_2, in a population of size N and that animals are killed in proportion to the rate at which they are encountered. Then, the rate of change of population size with time is:

$$dN/dt = -k(\alpha_1 Np + \alpha_2 Nq)$$

and the proportional rate of change in population size per unit of time is:

$$(dN/dt)/N = -k(\alpha_1 p + \alpha_2 q).$$

This is the real rate of proportional decline.

The apparent rate of decline from encounter rates ($S = \alpha_1 Np + \alpha_2 Nq$) is:

$$dS/dt = \alpha_1 d(Np)/dt + \alpha_2 d(Nq)/dt = \alpha_1(-k\alpha_1 Np) + \alpha_2(-k\alpha_2 Nq) = -kN(\alpha_1^2 p + \alpha_2^2 q).$$

Proportionally, this is:

$$(dS/dt)/S = -kN(\alpha_1^2 p + \alpha_2^2 q)/(\alpha_1 Np + \alpha_2 Nq) = -k(\alpha_1^2 p + \alpha_2^2 q)/(\alpha_1 p + \alpha_2 q).$$

Then the 'paradox coefficient' (PC), i.e. the ratio of the proportional change in encounter rates over the proportional change in population size, is:

$$PC = [(dS/dt)/S]/[(dN/dt)/N] = (\alpha_1^2 p + \alpha_2^2 q)/(\alpha_1 p + \alpha_2 q)^2.$$

By differentiating PC with respect to p and setting the derivative to zero, we find that the maximum paradox coefficient for a given α_1/α_2 occurs when $p = 1/(1 + \alpha_1/\alpha_2)$ and has maximum value $PC = (1 + \alpha_1/\alpha_2)^2/(4\alpha_1/\alpha_2)$.

Then, the maximum paradox coefficient in the case of the movement difference of the sperm whale clans and whaling vessels moving at 2 knots for which $\alpha_1/\alpha_2 = 1.06$ (see Table 9.8) is 1.0008, a tiny difference between the decline in sighting rate and decline in population size.

This does not mean that behavioural differences could not explain the sperm whale paradox, as there could be other behavioural causes of differences in sighting rates between groups of whales. However, behavioural heterogeneity has to be very large to explain the paradox entirely. To get a decline in sighting rates that is twice as large as the decline in population size, one needs differential sighting rates of $\alpha_1/\alpha_2 > 5.8$ (from max $PC = (1+\alpha_1/\alpha_2)^2/(4\alpha_1/\alpha_2)$). This is huge.

Acknowledgements

The authors are grateful to John Bannister and Karin Gleiter for access to data they had prepared for previous analyses; the Wildlife Conservation Society for access to the Townsend Abstracts; the National Weather Service of the US National Oceanographic and Atmospheric Administration for access to processed data from the Maury Abstracts; and the Blunt-White Library of the Mystic Seaport for providing access to the *Whalemen's Shipping List*. In addition, we are grateful to Steve Brennan, Frank Capitanio, Robert Gee and Steve Junker for careful extraction of data from 19th-century logbooks and other sources. This research was supported, in part, by the Alfred P. Sloan Foundation through the History of Marine Animal Populations programme, and in part by the Office of Ocean Exploration and the Northeast Fisheries Science Center, both of the National Marine Fisheries Service of the US National Oceanic and Atmospheric Administration. Finally, we are indebted to the New Bedford Whaling Museum and the Mystic Seaport for access to 19th-century whaling logbooks.

References

Allison, C. and Smith, T. D. (2004) *Progress on the Construction of a Comprehensive Database of Twentieth Century Whaling Catches*, Meeting Document SC/56/O 39, Scientific Committee, International Whaling Commission, Cambridge, UK

Bannister, J. L., Taylor, S. and Sutherland, H. (1981) 'Logbook records of 19th century American sperm whaling: A report on the 12-month project, 1978–79', *Report of the International Whaling Commission*, vol 31, pp821–833

Best, P. B. (1979) 'Social organization in sperm whales, *Physeter macrocephalus*', in H. E. Winn and B. L. Olla (eds) *Behavior of Marine Animals*, vol 3, Plenum, New York, pp227–289

Best, P. B. (1983) 'Sperm whale stock assessments and the relevance of historical whaling records', *Report of the International Whaling Commission*, Special Issue, vol 5, pp41–55

Best, P. B (2005) *Estimating the Landed Catch of Sperm Whales in the Nineteenth Century*, Meeting Document SC/56/IA5, Scientific Committee, International Whaling Commission, Cambridge, UK

Davis, L. E., Gallman, R. E. and Gleiter, K. (1997) *In Pursuit of Leviathan: Technology, Institutions, Productivity, and Profits in American Whaling, 1816–1906*, University of Chicago Press, Chicago, IL

Gregr, E. J., Nichol, L., Ford, J. K .B., Ellis, G. and Trites, A. W. (2000) 'Migration and population structure of northeastern Pacific whales off coastal British Columbia: An analysis of commercial whaling records from 1908–1967', *Marine Mammal Science*, vol 16, pp699–727

Hegarty, R. H. (1959) *Returns of Whaling Vessels Sailing from American Ports: A Continuation of Alexander Starbuck's 'History of the American Whale Fishery' 1876–1928*, Old Dartmouth Historical Society, and Whaling Museum, New Bedford, MA

Hope, P. L. and Whitehead, H. (1991) 'Sperm whales off the Galápagos Islands from 1830–50 and comparisons with modern studies', *Report of the International Whaling Commission*, vol 41, pp273–286

Kasuya, T. and Miyashita, T. (1988) 'Distribution of sperm whale stocks in the North Pacific', *Scientific Reports of the Whales Research Institute*, vol 39, pp31–75

Lund, J. N. (2001) *Whaling Masters and Whaling Voyages Sailing from American Ports: A Compilation of Sources*, New Bedford Whaling Museum, New Bedford, MA, Kendall Whaling Museum, Sharon, MA, and Ten Pound Island Book Co., Gloucester, MA.

Maury, M. F. (1852 *et seq*) *Whale Chart of the World* (The Wind and Current Charts), Series F, Sheet 2 (NW Pacific), Washington, DC

Maury, M. F. (1854) *Explanations and Sailing Directions to Accompany the Wind and Current Charts*, 6th edition, EC & Bros, New York, NY

May, R. M., Beddington, J. R., Clark, C. W., Holt, S. J. and Laws, R. M. (1979) 'Management of multispecies fisheries', *Science*, vol 205, pp267–277

Melville, H. (1851) *Moby-Dick*, Harper and Brothers, New York, NY

Punt, A., Friday, N. and Smith, T. D. (2007) 'Reconciling data on the trends and abundance of North Atlantic humpback whales within a population modeling framework', *Journal of Cetacean Research and Management*, vol 8, pp145–160

Rendell, L. and Whitehead, H. (2003) 'Vocal clans in sperm whales (*Physeter macrocephalus*)', *Proceedings of the Royal Society of London B*, vol 270, pp225–231

Sherman, S. C., Downey, J. M., Adams, V. M. and Pasternack, H. (1986) *Whaling Logbooks and Journals 1613-1927: An Inventory of Manuscript Records in Public Collections*, Garland Publishers, New York and London

Starbuck, A. (1878) 'History of the American whale fishery from its earliest inception to the year 1876', *Report of the US Fish Commission*, vol 4, 1875–1876, Appendix A

Tillman, M. F. and Breiwick, J. M. (1983) 'Estimates of abundance for the western North Pacific sperm whale based upon historical whaling records', *Report of the International Whaling Commission*, Special Issue, vol 5, pp257–269

Townsend, C. H. (1935) 'The distribution of certain whales as shown by logbook records of American whaleships', *Zoologica*, vol 19, pp1–50

Whitehead, H. (1995) 'Status of Pacific sperm whale stocks before modern whaling', *Report of the International Whaling Commission*, vol 45, pp407–412

Whitehead, H. (2002) 'Estimates of the current global population size and historical trajectory for sperm whales', *Marine Ecology Progress Series*, vol 242, pp295–304

Whitehead, H. and Rendell, L. (2004) 'Movements, habitat use and feeding success of cultural clans of South Pacific sperm whales', *Journal of Animal Ecology*, vol 73, pp190–196

Wood, D. (1831–1873) *Abstracts of Whaling Voyages*, 5 vols, New Bedford Whaling Museum, MA

WSL (*Whalemen's Shipping List and Merchants' Transcript*) (1843–1914) vols 1–72, 17 March 1843–29 December 1914, New Bedford, MA

10

Understanding the Dynamics of Fisheries and Fish Populations: Historical Approaches from the 19th Century

Julia Lajus

The idea that historical sources can significantly enrich our understanding of marine animal population dynamics has gained recognition in recent years. This is not an altogether new approach, however, for during the 19th century a number of scholars produced studies that would now be cast in the mould of historical ecology. Although their works were few in number, they warrant investigation because of their contemporary novelty – none were fully accepted by the scientific community of their day – and their relevance to the current surge of interest in holistic approaches to fisheries management.

This chapter examines the contributions of Russian and Scandinavian scientists who pioneered the notion that history and natural science could interact to enhance knowledge and understanding of the ocean realm. It focuses on the ideas and legacy of Karl Ernst von Baer (1792–1876), a zoologist of Baltic German origin who worked at the Imperial Russian Academy of Sciences in St Petersburg for a significant part of his career. Attention is also afforded to the interdisciplinary approaches adopted by Jonas Axel Boeck (1833–1873) and Axel Ljungman (1841–1901), who were based in Norway and Sweden, respectively, and whose work developed independently from that of Baer during the second half of the 19th century. Two types of historical evidence underpin this appraisal. First, the published works of these researchers are used to demonstrate their interdisciplinary perspectives on the dynamics of fish populations, many of which – notably Baer's conceptual paper on this theme (Baer, 1854) – were not widely read by contemporary colleagues, partly because they were not published in English, a fact which still obscures their significance today. Second, the archival material generated by Baer and housed in St Petersburg (PFA RAN) is deployed to illustrate and appraise his innovative approach to the study of fisheries. The main point to emerge from this investigation is that analysis of past trends can help scientists and managers to address current issues, but biologists and historians – then and now – are instinctively wary of pursuing this course.

BAER'S INVESTIGATIONS INTO FISHERIES HISTORY

Karl Ernst von Baer was an eminent zoologist. One of the founders of modern embryology, he is primarily known for the discovery of the mammalian ovum (Raikov, 1968; Lenoir, 1989). Very well educated and an open-minded naturalist, Baer had a great interest in the history of both the natural and human worlds, and referred to history, in a very broad sense, as his vocation. Baer also treated embryology as a history: the history of development, or *Entwicklungsgeschichte* in German. In his later days, when he became more concerned about human social problems, Baer undertook research in a number of disciplines closely related to history, such as historical geography, ethnography and archaeology. According to one of his biographers:

> *Baer was not wrong, calling himself a historian; an aspiration for the historical perception of any subject clearly outlines all von Baer's scientific studies; studying any subject, from any possible point of view, he is not satisfied by learning the present of this subject, he moves further trying to fathom its depth, to learn how this present status arose and thus enters the field of history; for Baer as for many other great naturalists of our century, the natural history is a history in the full meaning of this word.* (Kuznetsov, 1892, pp43–44)

His deep interest in the processes of development in both nature and human societies led him to advocate the relevance of the natural sciences to socio-economic problems (Baer, 1986). This perspective, which first emerged during Baer's travels to the north of Russia in 1837 and 1840, came from his work with fisheries. Prior to these trips and upon his return, Baer collected materials on the history of the region, including some trade and fisheries statistics (PFA RAN[1]). Thus, it was not by chance that Baer was appointed to head the first Expedition for the Study of the Status of Fisheries in Russia in 1851, which was organized by the Ministry of the State Domain and tasked to focus on Lake Peipus and the Russian coasts of the Baltic Sea (Baer, 1860). The expedition was mounted in response to the first signs of trouble in the Russian fisheries. In 1850, local governors of the Baltic provinces reported to the central authorities that a substantial decline in fish catches had occurred. Baltic herring and the small freshwater smelt were the main sources of food for coastal populations; therefore, the simultaneous contraction in landings of both species led to serious social problems. Baer was asked by the government to determine the cause of this decline and to provide a forecast for the future of these fisheries. His response was to propose a scientific programme for collecting data from archives and private fishery enterprises (Lajus, 2002).

Baer began his studies by obtaining catch statistics and examining fishing techniques and seasons. He tried to collect catch data for previous years in order to assess the extent of the decline in the fisheries and to evaluate whether the activity should be officially regulated. Baer corresponded with local officials and the aristocrats who owned manors on the shores of the Baltic in an attempt to obtain information on catches. He also searched in archives for old documents relating to the fish trade and fisheries (Lajus et al, in press). He stressed the necessity of obtaining reliable quantitative data that stretched as far back in time as possible, emphasizing how:

> ... *the often repeated complaint that catches are becoming smaller and smaller one cannot take into account, because always when fisheries take place on large scale, the fabulous stories on the enormous catches in previous years are very often recognized as the normal catches.* (Baer, 1854, p481)

Unfortunately, the catch statistics kept by these manors were very poor. The longest time series Baer was able to obtain amounted to 50 years for the herring and several other fish species in the manor of Grossenhoff (now Suuremõisa) on the island of Dago (now Hiiumaa). Yet, this time series probably constitutes the longest available for scientists examining the Baltic Sea region (Lajus et al, in press).

In addition to collecting data on catches from the owners of manors, Baer undertook some research in the archives at Revel (now Tallinn) and Narva, which 'were open to him with a great readiness' (Baer, 1860, p13). He noted that documents in these archives 'contain some hints of the quantity of catches in the previous centuries and of the development of the fish trade, which has some significance'. But, unfortunately, in Baer's opinion:

> *The magistrates [of these towns] were more concerned with keeping privileges for fisheries in certain waters for the town itself or its separate communities, or with the supply of town with fish from the other regions, or, at last, with the good quality of fish on the market, than with the conservation of fish in the waters which were under their authorities.* (Baer, 1860, p13)

Only a few notes from Baer's work in these archives were found in his papers at the St Petersburg Branch of the Archives of the Russian Academy of Sciences (PFA RAN) where most of the materials relating to his studies of fisheries in the 1850s are kept. The Baltic part of this collection has not been well studied, and even Baer's expedition diary has not been transcribed and interpreted. The information he obtained from archives in Tallinn and Narva includes data on the herring trade, both quantities and prices, handled by these towns for a number of years during the first half of the 19th century (PFA RAN[2]).

Baer also tried to collect information on taxes imposed upon the fisheries. The earliest was dated 1572, revealing that 'the miller Joh. Nykerk of the Estate of Schloss Fall had been paid with salmon, lamprey, whitefish and bream instead of money' (PFA RAN[3]). Moreover, 'in 1611, Hinrick Kursell was allowed to catch fish at the coast of the estate of Kiwidepah [now Kiideva, on the west coast of Estonia] on the condition that he would give 80,000 Baltic herrings to the estate' (PFA RAN[4]). It was also recorded how 'from 1727 to 1807, Sweden fishermen from Kertell [on the Dago island], when they caught fish in Matzal-Wieck [now Matsalu] Bay, gave to the pastor ... in spring 1000 perches or breams per boat and in autumn one quarter barrel of salted pike or ide per boat' (PFA RAN[5]). But this, together with related information, was too fragmentary to facilitate analysis, although it provided some evidence as to the scale of fisheries and target species in earlier periods. In an attempt to better understand the organization of fisheries, Baer studied the state charters of Lievland (now Latvia) for evidence of the fishing rights that prevailed before the mid 19th century. Copies and notes from these charters were collected by Baer's assistant, Alexander Schulz (PFA RAN[6]), who also researched documents pertaining to Revel (Tallinn) town council (Baer, 1860, p22).

Baer found that salmon fisheries had existed at Narva since the foundation of the town during the 13th century (Baer, 1854, p480). He also compared old charts of the River of Narva – dated 1695, 1743, 1744 and 1756 – with more recent maps in order to understand the decline of herring fisheries in Narva Bay (Baer, 1860, p31). Baer proposed that this decline was caused by decreasing salinity in the western part of the bay and adjacent marine waters due to the change of the river current as a result of the destruction of the dams constructed in 1765 to 1767 (Baer, 1860, p31; maps: PFA RAN[7]). In addition, Baer's archives in St Petersburg include several folders containing descriptive materials regarding the history of fisheries in the Baltic Sea, the Neva, Volkhov and Narva rivers, and Ladoga Lake (PFA RAN[8]). He also noticed a drastic decline of sturgeon in the Baltic Sea, but did not provide any historical data on its earlier abundance (Baer, 1854, p470).

Baer collected material on the history of fisheries in Sweden in connection with his trip there in 1852. He was very much interested in the Swedish experience, especially with regard to the fisheries regulations that had recently been instituted (PFA RAN[9]). He paid special attention to the history of the old and highly erratic herring fisheries in the Bohuslan region, collecting excerpts from the reports and papers of Sven Nilsson (1787–1883), a well known Swedish zoologist (Regnell, 1987) who was among the first scientists to describe the Bohuslan fisheries and investigate why they fluctuated so dramatically (PFA RAN[10]).

It is important to note that the task of the Baltic expedition under Baer's leadership was very precise – namely, to understand the causes of decline of herring fisheries and to propose measures for their improvement. Baer recognized that most herring fisheries did not decline continuously, but only experienced fluctuations caused by natural forces. This, in his opinion, was mostly related to weather conditions affecting spring fisheries. Thereafter, he concentrated on the Lake Peipus fishery, which was in real decline and required the introduction of special legislation. Yet, historical research into this subject was rendered impossible by a fire in the archives of Pechersky monastery before Baer arrived (Baer, 1854, p482). In any case, it would seem that he did not have enough time to undertake historical studies of the Baltic fisheries. He was therefore unable to publish the information he had found, although the information gathered from the owners of manors was published in summary form (Baer, 1860). The detailed data contained in his papers have only become available recently (Lajus et al, in press).

The Russian authorities were so impressed with Baer's studies that they entrusted him with the organization of a similar expedition to the Caspian Sea (1853–1856), which had supported the nation's most important fisheries. This was followed by investigations into other regional fisheries, notably those conducted in the White and Barents Sea (1859), the Azov and Black Sea (1862–1864), and a number of lake fisheries in northwestern Russia (1867–1870). After 1856, the expeditions were led by Baer's assistant on the Caspian expedition, Nikolay Danilevsky; but Baer remained interested in the results. In effect, his Baltic expedition had initiated a research programme that lasted for 20 years and involved the systematic collection of data on fisheries conducted in various bodies of water. This scientific survey, which was probably the first to be undertaken on a national scale (Smith, 1994, p35), produced nine thick volumes describing the fisheries, with four volumes of colour plates detailing the construction of fishing vessels and gear.

Baer was among the first scientists to collect fisheries statistics. Gathering all the available data relating to earlier periods and establishing the need to collect statistics on a regular basis were two of his major achievements. In practice, the descriptions of fishing activity prepared by Baer and his colleagues – together with the statistical data – served as a starting point for analysing the development of Russia's fisheries in subsequent years. While the provision of such baseline material was one of Baer's major contributions, he also made important recommendations for the improvement of legislation and offered informed advice to fishers.

When the area of study expanded to the Caspian, White and Barents seas, collecting historical data remained an important part of the work. For instance, Baer's assistant and successor, Danilevsky, conducted research in the private archives of the owners of the Caspian fisheries (Lukina, 1984, pp404, 420), an impressive feat given that few owners allowed scientists to access their records (Lukina, 1984, p424). Studies published in the expedition report contain invaluable statistical material on the landings of the Transcaucasian Caspian fisheries (on the Kura River and adjacent sea coasts) from 1829 to 1846 (Danilevsky, 1863). They comprise the daily catch data reported by 43 fishing teams over eight years. In Baer's opinion, these data had additional value because they appraised the fisheries in a region which had only just become part of the Russian Empire. As large-scale commercial fishing had only just started, 'for the first time the complete history of the enormous fishery, which does not exist in any literature, will be compiled' (Baer, 1861, p133). While statistical data preceding the respective expeditions by 20 or 30 years could not be treated as 'historical', it proved difficult to find earlier data for most of the fisheries studied by Baer and his colleagues. Yet, their work remains distinguished by the effort invested in the discovery of early data and attempts to make it more widely available.

In some cases, however, data from an early period were uncovered. For example, Danilevsky, when leading the White Sea expedition, obtained data from the Arkhangelsk customs archive relating to the delivery of cod to this port from the Barents Sea during the 1782 to 1790 period. Following analysis of this material, he rejected the common opinion that the cod fisheries had been more productive during the 18th century than they were in the 1850s (Danilevsky, 1862, pp142–143). This question arose again on the eve of the 20th century when a new set of studies into the Russian fisheries commenced under the leadership of Nikolay Knipowitsch. New archival research, influenced by previous studies of Baer and Danilevsky, showed that the port of Arkhangelsk received less fish from the Barents Sea cod fisheries between 1762 and 1763 than it did during the late 19th century (Romanov, 1902).

In 1853, Karl Baer presented his findings on the history of Russia's fisheries to the Academy of Sciences. Subsequently, an extended version of this presentation was published in German and in Russian under a title that translated as 'Materials for the history of fisheries in Russia and in the seas belonging to it' (Baer, 1853, 1854). In this innovative work he advocated the use of historical data for understanding changes in fisheries and the status of fish populations, as well as proposing further avenues of research. As Baer pointed out: 'in the historical writings fisheries usually are not mentioned at all: the subject looks too insignificant.... Even in the description of customs and modes of people's food supply ... there is only very scarce information about fisheries' (Baer, 1854, p466). For what purpose, besides pure historical interest, he asked rhetorically, is it necessary to collect information on the history of fisheries? His reply appears to reflect modern thinking: 'only on the base of the thorough historical studies would it be possible to judge what kinds of fish species ... are [being caught] now in smaller amounts than earlier' (Baer, 1854, p470). It is important to note that Baer clearly recognized the necessity of determining changes in fishing effort from historical documents, as well as the size of catches. He formulated the question thus: 'what was the development of fisheries regarding the increase of human population, the successes of industry, the increase and acceleration of transportation?' (Baer, 1854, p469).

In addition to catch data, Baer was interested in the development of fishing gear, the organization of fisheries and, in particular, the evolution of fisheries legislation. He did not confine himself to Russia when enquiring into such matters, frequently offering comparisons across different cultures, thereby placing Russia in the broader context of European history. The comparisons between Russian and German fisheries are especially interesting. Baer emphasized the rationality of ancient German fishing regulations in fisheries, which, for example, stipulated that a gap should be left in the centre of weirs for spawning fish to pass through. In contrast, he castigated the Russians for their prodigal, improvident attitude towards fish resources, although he noted that they possessed an enterprising spirit in comparison, for instance, with Estonian fishers.

In his search for historical accounts of fishing activities, Baer relied heavily upon classical historical works, including the first comprehensive history of Russia by Karamzin (1816–1829), as well as histories of Northern Europe by Schlözer (1771), Prussia by Voigt (1827–1836), and Norway by Torfeaus (1711) and Pontoppidan (1754), to name but a few. He widely referred to more specialized publications containing information on the development of fisheries, such as Bock (1769), Ekström (1835), Noël (1800), Faber (1829) and Nilsson (1828). This notwithstanding, Baer's studies on the history of fisheries are not widely known, largely because they were published in Russia.

Baer had hoped to publish a more comprehensive study of the history of Russian fisheries based on information collected from chronicles, decrees and monastic records. 'It is evident', wrote Baer, 'that the whole set of data is much more instructive than the fragmentary figures. Pulling them together, it might be possible to get a clear idea of the development of fisheries in the given part of Russia' (Baer, 1854, p482). He expected the archives of monasteries to be especially rich in data on annual catches (Baer, 1854). In this, he was correct, as recent interdisciplinary studies based on the archives of monasteries situated in the Russian North have demonstrated (Lajus et al, 2001, 2005). Baer also attempted to collect excerpts from counting books with quantitative information about catch sizes, fish prices and the rents of fishing grounds. He called upon contemporary historians and local educated people to assist him in these endeavours by forwarding any relevant information that they could obtain. However, Baer was constrained by the fact that his request failed to generate a response.

SCANDINAVIAN HERRING STUDIES

The importance of the past in understanding strong fluctuations in fish stocks and fisheries was recognized at an early juncture in Scandinavia. In terms of the development of fish science, especially regarding herring, most of the 19th century could be termed the 'Scandinavian period' (Heincke, 1898). Interest in the highly volatile Bohuslan fishery is particularly apparent. In 1826, Nilsson, who was the first scientist to explore the reasons underlying the collapse of this fishery in 1808, proposed the far-reaching conclusion that herring stocks consist of numerous local populations. His findings became widely known and subsequent work on the causes of these fluctuations, especially by the Danish zoologist Henrik Kröyer, generated important work on the subspecies of herring by Friedrich Heincke (Sinclair and Solemdal 1998; Smith, 1994, p22; Jansen, 2002). Heincke was disinclined to take a historical approach towards vacillating fish abundance and his contribution was purely biological. However, he did devote a chapter of his *magnus opus* on herring to the history of herring studies (Heincke, 1898), in which he cited the works of Axel Boeck from Norway and Axel Ljungman from Sweden, scientists who had both applied historical data to their studies of the herring fisheries.

In 1859, the Norwegian government funded scientific studies of marine fisheries for the first time. Jonas Axel Boeck (1833–1873) was tasked with researching a complex of problems arising from the herring fisheries (Solhaug and Saetersdal, 1971/1972, p401; Sakshaug and Mosby, 1976). His mandate was both historical and scientific. In his attempts to chart herring periodicity and connections between catches of herring in different areas, Boeck used archival data (Schwach, 2000, pp32–37). In the judgement of modern scientists, his work 'suffered from this necessity to devote himself both to the history of fisheries and to practical fishery research' (Solhaug and Saetersdal 1971/1972, p401). Boeck's main work, entitled *On Herring and the Herring Fisheries*, was published in 1871, while his book on the history of the Norwegian herring fishery remained unfinished due to his early death (Boeck, 1871, 1873).

Boeck's studies were continued by Axel Ljungman (Anonymous, 1906), a Swede who made a new attempt to understand the fluctuations of the Bohuslan herring fishery. Like Baer and Boeck, it was the study of herring, one of the most important commercial species in Northern Europe, which encouraged Ljungman to use historical sources to understand fluctuations in the fisheries. In his major paper on this subject, Ljungman (1879) coined the term 'herring period' and devised a list of such phases that dated back to the year 1000 (Lindquist, 2000, p345). However, he was careful to state that 'from the time previous to the year 1300 we have scarcely any information regarding our Bohuslan herring fisheries' (Ljungman, 1882, p500). Ljungman's relation of the incidence of herring appearance to solar cycles provided the foundation for studies in the periodicity of natural phenomena. His ideas concerning the utility of historical data in natural sciences were particularly progressive and entailed a programme for further studies of the problem that included both the systematic deployment of historical evidence to determine periodicity and the recording of long-term observations of natural conditions, such as hydrology and climate, which might have impacted upon herring shoals (Lajus, 2005, p105). While the second half of this programme started to develop soon afterwards (but not in the way Ljungman had proposed), the initiation of historical studies was postponed until more recent times.

Using qualitative evidence, Ljungman reviewed the causes of the appearance and disappearance of herring, dividing them into three distinctive groups: unexplained or accidental; caused by human agencies; or due to natural causes (Ljungman,

1882, pp498–499). Calling for the use of historical sources in fisheries research, Ljungman wrote:

> ... it cannot be expected of scientists that for the sake of a few data they should go through the immense collections of documents in our archives or through the whole range of historical literature, and thus gather all the necessary facts from former times; but these facts should be obtained by persons who made a life work of the study of history. By such work done by the officers of the Royal Norwegian Archive, it became possible for Axel Boeck to furnish quite a number of new and interesting contributions to the history of Bohuslan fisheries. (Ljungman, 1882, p502)

On the basis of Boeck's experience, Ljungman (1882, p502) proposed that 'scientists and historians should work hand in hand'.

Conclusion

More than a century and a half has passed since Baer first assembled historical material in an attempt to understand the dynamics of fisheries and fish populations. It has been more than 100 years since Boeck and Ljungman independently adopted the same approach in another part of Northern Europe. Baer himself understood very well that the work he called for would not be easy, and both biologists and historians would not accept his message immediately. He also saw that time was needed to launch such studies and that even more time would be required to bring them to fruition:

> I know very well how incomplete my data are, but I also know that they will not reach the necessary completeness in all my lifetime. If even I would be able to turn the overall attention towards the history of fisheries in our motherhood, even then it might be that two centuries will pass before all the extant data would be sampled. (Baer, 1854, p466)

Indeed, none of the historical studies described in this chapter exerted a significant influence upon contemporaries. During the late 19th century, some 50 years after the publication of Baer's 1853/1854 paper, the author of one of the major Russian works on fisheries wrote: 'unfortunately, Baer's hope was not realized and the task that he posed remains unfulfilled by now and the question is open for the future researchers in the history of fisheries' (Veshniakov, 1894, p299). Increased awareness of, and concern about, the fate of the environment and natural resources have recently influenced the development of both history and science. Accordingly, the value of an interdisciplinary approach has become much more apparent and biologists have followed Ljungman's advice and started to work 'hand in hand' with historians. Although this has taken more time than Baer would have liked, his pioneering work is considerably more relevant now than in the period it was written.

ACKNOWLEDGEMENTS

I would like to thank Brian McKenzie for drawing my attention to the work and legacy of Karl Ernst von Baer, Tim Smith for his encouragement and Artur Svansson for generously providing me with biographical information on Swedish scientists. I am also grateful to the two anonymous referees who supplied helpful comments on a draft of this chapter.

REFERENCES

Anonymous (1906) 'Ljungman, Axel Wilhelm', in H. Hofberg, F. Heurlin, V. Millqvist and O. Rubenson (eds) *Svenskt biografiskt handelexikon*, Albert Bonniers Förlag, Stockholm, p87

Baer, K. E. von (1853) 'Materialen zu einer Geschichte des Fischfanges in Russland und den angranzenden Meeren', *Bulletin scientifique physico-mathématique de l'Académie Impériale des Sciences de St.-Pétersbourg et rédigé par le secrétaire perpétuel*, vol XI, N15–18, pp225–254, 257–288

Baer, K. E. von (1854) 'Materialy dlia istorii rybolovstva v Rossii i v prinadlezhaschikh ei Moriakh' ['Materials for the history of fisheries in Russia and the seas belonging to it'], in *Zapiski Imperatorskoi Akademii Nauk po I i III otdeleniiu*, vol 2, vyp 4, pp465–544 (in Russian)

Baer, K. E. von (1860) 'Rybolovstvo v Chudskom i Pskovskom Ozerakh i v Baltiiskom More' ['Fisheries in the Chudskoe and Pskovskoe lakes and in the Baltic Sea'] in *Issledovaniia Rybolovstva v Rossii*, vol I, Ministerstvo Gosudarstvennykh Imuschestv, St Petersburg (in Russian)

Baer, K. E., von (1861) 'Rybolovstvo v Kaspiiskom More', in *Issledovaniia Rybolovstva v Rossii*, vol II, Ministerstvo Gosudarstvennykh Imuschestv, St Petersburg (in Russian)

Baer, K. E. von (1986) *Autobiography*, [translation of 2nd edition (1886), J. Oppenheimer (ed), H. Schneider (translator)], Science History Publications, Canton, MA

Bock, F. S. (1769) *Versuch Einer Vollständigen Natur- und Handlungsgeschichte der Heringe*, Königsberg

Boeck, A. (1871) 'Om silden og sildefiskerierne navnlig om det Norske Vaarsildfiske', in *I Indberetning til den kgl. Norske Regjerings Departement for det Indre om foretagne praktisk-videnskabelige Undersøgelser*, Trykt efter Foranstalting af det kgl. Departement for det Indre, Christiania (in Norwegian)

Boeck, A. (1873) 'Det Bohuslenske Sildefiskeries Histori', *Nordisk Tidskrift for Fiskeri*, 1 Aarg., pp1–27

Danilevsky, N. Ya. (1862) 'Rybnye i Zverinye Promysly v Belom i Ledovitom Moriakh' ['Fisheries and marine hunting in the White Sea and the Arctic Ocean'], in *Issledovaniia o Sostoianii Rybolovstva v Rossii*, vol VI, Ministerstvo Gosudarstvennykh Imuschestv, St. Petersburg (in Russian)

Danilevsky, N. Ya. (1863) 'Tablitsy Ulovov po Zakavkazskomu Rybolovstvu za vremia Kazennogo Upravleniia s 1829 po 1846 god i za vremia Otkupnogo Soderzhaniia ikh Pochetnym Grazhdaninom Arshakuni s iiulia 1847 po iiul 1855 goda' ['Tables of catches in Transcaucasian fisheries during the time of their belonging to the Treasury and the time of their renting by the Honorary Citizen Arshakuni from July 1847 to July 1855'], in *Issledovaniia o Sostoianii Rybolovstva v Rossii*, vol V, Ministerstvo Gosudarstvennykh Imuschestv, St Petersburg, pp1–130 (in Russian)

Ekström, C. U. (1835) *Die Fische in den Scheeren von Mörkö* (Aus Swedishen ubersetz. Dr. Creplin), G. Reime, Berlin

Faber, F. (1829) *Naturgeschichte des Fische Islands (mit einem Anhange von den Isländischen Medusen und Strahltieren)*, Brönner, Frankfurt am Main

Heincke, F. (1898) *Naturgeschichte des Herings* (Abhandlungen des Deutschen Seefischerei-Vereins, Bd. 2), Berlin

Jansen, S. (2002) 'Den Heringen einen Paß ausstellen. Formalisierung und Genauigkeit in den Ahfängen der Populationsökologie um 1900', *Berichte zur Wissenschaftsgeschichte*, vol 25, pp153–169

Karamzin, N. M. (1816–1829) *Istoriia Gosudarstva Rossiiskogo* [*History of the Russian State*], 12 vols, Tipografiia Glavnogo shtaba, St Petersburg (in Russian)

Kuznetsov, I. D. (1892) 'Akademik Karl Ernst (Karl Maksimovich) von Baer, ego zhizn i deiatelnost, preimuschestvenno v oblasti ikhtiologii – nauchnoi i prikladnoi' ['Academician Karl Ernst von Baer (Karl Maksimovich), his life and activities, especially in the field of scientific and applied ichthyology'], *Vestnik Rybopromyshlennosti*, vol 12, pp1–72 (in Russian)

Lajus, D. L., Dmitrieva, Z. V., Kraikovski, A. V., Lajus J. A., Yurchenko A. Y. and Alexandrov, D. A. (2005) 'Use of historical data on catches of commercial fish species for the tracing of the influence of climate on fish populations: Cases of the White and Barents sea fisheries in 17th–18th centuries', *ICES Journal of Marine Sciences*, vol 62, pp1426–1435

Lajus, J. A. (2002) 'Karl Baer kak istorik rybolovstva' ['Karl Baer as historian of fisheries'], in *Materialy nauchnoi konferentsii 'Istoricheskii opyt nauchno-promyslovykh issledovanii v Rossii'*, VNIRO, Moscow, pp135–136 (in Russian)

Lajus, J. A. (2005) 'Influence of weather and climate on fisheries: Overview of the early history of the problem, 1850s–1950s', in S. Emeis and C. Lüdecke (eds) *From Beaufort to Bjerknes and Beyond: Critical Perspectives on Observing, Analyzing and Predicting Weather and Climate*, Rauner Verlag, Augsburg

Lajus, J. A., Alekseeva, Y. A, Davydov, R. Dmitrieva, A., Kraikovsky, Z. V., Lajus, A. V., Lapin D. L., Mokievsky V. V., Yurchenko, V. O., Alexandrov, D. A. (2001) 'Status and potential of historical and ecological studies on Russian fisheries in the White and Barents seas: The case of the Atlantic salmon (*Salmo salar*)', in P. Holm, T. D. Smith and D. J. Starkey (eds) *The Exploited Seas: New Directions for Marine Environmental History*, International Maritime Economic History Association, St John's, Newfoundland, no 21

Lajus, J., Ojaveer, H. and Tammiksaar, E. (in press) 'Fisheries on the Estonian coast of the Baltic Sea in the first half of the 19th century: What can be learnt from the archives of Karl Ernst von Baer', *Fisheries Research*

Lenoir, T. (1989) *The Strategy of Life: Teleology and Mechanics in Nineteenth Century German Biology*, University of Chicago Press, Chicago, IL

Lindquist, A. (2002) 'Herring periods of Bohuslan: A cross-sectoral approach', *ICES Marine Science Symposia*, vol 215, pp343–351

Ljungman, A. V. (1879) 'Bidrag till lösningen af frågan om de stora sillfiskenas sekulära priodicitet', *Tidskrift för Fiskeri*, vol 5, pp257–268 (in Swedish)

Ljungman, A. (1882) 'Contribution towards solving the question of the secular periodicity of the great herring fisheries', *US Commission of Fish and Fisheries*, Washington, DC, VII, pp497–503

Lukina, T. A. (1984) *Kaspiiskaia Ekspeditisiia K. M. Baer'a. 1853–1857 gg. Dnevniki i Materialy* [*The Caspian Expedition of Karl Baer. 1853–1857. Diaries and Materials*], Nauka, Leningrad (in Russian)

Nilsson, S. (1828) *Fornyad Berattelse om Fiskerierna i Bohus-Lan*, Lundae

Noël de la Moriniére, S. B. (1800) *Tableau Historique de la Peche de la Baleine*, Paris

PFA RAN[1] (St Petersburg Branch of the Archives of the Russian Academy of Sciences) Karl Ernst von Baer Papers, Collection (f.) 129, inventory (op.) 1, files (dd) 313, 344, 360, 415–417 and others

PFA RAN[2] f. 129, op. 1, d. 473, ll. 10–16

PFA RAN[3] f. 129, op. 1, d. 499, l. 107

PFA RAN[4] f. 129, op. 1, d. 499, l. 107

PFA RAN[5] f. 129, op. 1, d. 499, l. 107

PFA FAN[6] f. 129, op. 1, d. 483

PFA RAN[7] f. 129, op. 1, dd. 479–481

PFA RAN[8] f. 129, op. 1, dd. 465, 494–498

PFA RAN[9] f. 129, op. 1, d. 444

PFA RAN[10] f. 129, op. 1, dd. 441–443

Pontoppidan, E. (1754) *Versuche einer Natürlichen Historie von Geschichte Norwegens*, Copenhagen, Denmark

Raikov, B. E. (1968) *Karl Ernst von Baer 1792–1876: Sein Leben und sein Werk*, Übersetzt von H. v. Knorre, Leipzig, Germany

Regnell, G. (1987) 'Sven Nilsson, 8/3 1787–30/11 1883', *Minnesteckning vid Kungl. Vetenskapsakademien*, vol 189 (in Swedish)

Romanov, N. V. (ed) (1902) 'Treskovyi promysel v polovine XVIII stoletiia' ['Cod fisheries in the mid 18th century'], in *Statisticheskoe issledovanie Murmana*, vol 1, no 1, pp1–16, Komitet dlia pomoschi pomoram Russkogo Severa, St Petersburg (in Russian)

Sakshaug, E. and Mosby, H. (1976) 'En oversikt over norsk oseanografis historie fram til den 2.verdenskrig' in E. Sakshaug, F. E. Dahl and N. P. Wedege (eds) *Norsk oseanografi – status og perspektiver*, Norsk Oseanografisk Komite, Oslo, pp209–232

Schlözer, A. L. von (1771) *Allgemeine Nordische Geschichte*, Halle

Schwach, V. (2000) *Havet, Fisken og Vitenskapen. Fra Fiskeriubdersøkelser til Havforskingsintitutt 1860-2000*, Havetforskning Institut, Bergen, Norway

Sinclair, M. and Solemdal, P. (1988) 'The development of "population thinking" in fisheries biology between 1878 and 1930', *Aquatic Living Resources*, vol 1, pp189–213

Smith, T. P. (1994) *Scaling Fisheries: The Science of Measuring the Effects of Fishing, 1855–1955*, Cambridge University Press, Cambridge, UK

Solhaug, T. and Saetersdal, G. (1971/1972) 'The development of fishery research in Norway in the nineteenth and twentieth centuries in the light of the history of the fisheries', *Proceedings of the Royal Society of Edinburgh, Section B (Biology)*, vol 73, pp399–412

Tammiksaar, E. (2000) 'Karl Ernst von Baer als Meeresbiologie', in J. Kaasch, M. Kaasch and R. Kinzelbach (eds) *Berichte zur Geschichte der Hydro – und Meeresbiologie und weitere Beitrage der 8. Jahrestagung der Deutschen Gesellschaft fur Geschichte und Theorie der Biologie,* pp111–117

Torfaeus, T. (1711) *Historia rerum Norwegicarum*, 4 vols, Havnia

Veshniakov, V. I. (1894) *Rybolovstvo i Zakonodatelstvo* [*Fisheries and Legislation*], Trenke & Fusno, St Petersburg (in Russian)

Voigt, J. (1827–1836) *Geschichte Preussens*, 9 vols, Königsberg

11
A Political History of Maximum Sustained Yield, 1945–1955

Carmel Finley

Although it remains at the heart of fisheries management policies, the concept of maximum sustained yield (MSY) has been heavily criticized by scientists for its failure to sustain global fish stocks (Larkin, 1997; Pauly, 1994). The most well-known example of commercial extinction is Atlantic cod (*Gadus morhua*); but this is by no means the only population to have come under threat. Scientists estimate that the number of large ocean predators has declined by as much as 90 per cent (Myers and Worm, 2003), with slow-growing, long-lived species, such as orange roughy (*Hoplostethus atlanticus)* and west coast rockfish (*Sebastes)*, which inhabit the deep ocean, being especially hard hit (Berkeley et al, 2004).

Much less attention, however, has been paid to MSY's role in the policy-making of nation states. The US played a pivotal role in this respect during the years that followed the end of World War II. Having adopted MSY as the principal goal of its fisheries management policy in 1949, the US endeavoured to persuade all other fishing nations to embrace MSY, a strategy that culminated in a conference sponsored by the United Nations (UN) and held in Rome in 1955. During this period, MSY formed an important facet of the US's Cold War foreign policy in that it helped to justify the contention that American ships, submarines, airplanes and fishing boats should have the freedom to use the seas. This entailed the freedom of vessels from the world's developed fishing nations to continue fishing off the shores of poorer countries, many of which objected vehemently to the presence of foreign distant-water fleets. The adoption of MSY at the Rome Conference meant that until countries could prove scientifically that stocks had fallen below prescribed levels, fishing could not be restricted or managed. One of the legacies of the 1955 decision was the establishment of a pattern that is still repeated in the early 21st century. In essence, fish are taken as fast as possible until critical biological thresholds are reached. At this point, measures are applied to slow or halt the fishing; but these are frequently difficult to implement and generally ineffective in curtailing catching activity sufficiently to allow stocks to recover productivity.

Scholars generally agree that American political power was at its zenith in the years immediately following World War II. But despite the unprecedented strength of the US navy, and the resources expended on the development of oceanography, this political power was not reflected in a successful fishing industry (Hollick, 1981). By the early 1950s, the US had ceded its place as the world's third leading fishing nation to the Soviet Union, and was importing much of its seafood. But proponents of this argument have only looked at the percentage of the global fish catch caught by Americans. A different picture emerges if the focus is placed on the projection of American power in the realm of science and the growing influence that this had upon foreign policy during the post-war years. Ronald E. Doel, for instance, argues that US scientists were increasingly influential in foreign policy issues after 1945, an expanded role that shaped the growth of scientific disciplines and institutions (Doel, 1997, p215). Likewise, John Krige contends that science 'was embedded in, and instrumentalized for, the projection of American power in post-war continental Europe', where the US 'attempted to use its scientific and technological leadership, in conjunction with its economic, military and industrial strength, to shape the research agendas, the institutions and the allegiances of scientists ... in line with US scientific, political, and ideological interests in the region'(Krige, 2006, p3).

Whereas Doel and Krige are concerned with American influence in the more traditional realms of earth science and physics, their arguments apply equally to a variety of environmental issues, especially fisheries. The adoption of MSY as the goal of US fisheries policy, and the manoeuvring to have it recognized at the international level, essentially shaped post-war fisheries science. The model of fisheries management preferred by the US precluded an international body to adjudicate fisheries conflicts in favour of limited regional, bilateral or multilateral agreements among countries, as embodied in the International Commission for North Atlantic Fisheries (ICNAF) and the Inter-American Tropical Tuna Commission (IATTC). The goals for both commissions, which were contoured by US foreign policy concerns, were established by policy-makers before discussions and negotiations among scientists could begin.

This chapter examines the political context in which these policy decisions were made. It contends that the rivalry between Japan and the US for the domination of the rich salmon resources of the North Pacific, and the high-seas tuna of the tropical Pacific, greatly influenced both fisheries science and fisheries management during this period. These issues coincided with the application by the fishing industry of various wartime technical innovations, leading to a growth in catching power that was greatly accelerated by the investment of large quantities of state money. As distant-water fisheries expanded, new conflicts over territorial claims emerged and fishing became increasingly intertwined with foreign policy objectives.

Conflict over fisheries in the Pacific

Analyses of the historical development of fisheries have tended to focus on the North Atlantic, largely because fishing activity in this region, especially the North Sea, has evolved over hundreds of years. In contrast, the development of commercial fishing in the Pacific has occurred over a much shorter time-frame. Although local and regional artisanal fisheries had long since been practised in areas of the Pacific, it was not until the 20th century that industrialized harvesting techniques were introduced, their deployment spreading rapidly as the century wore on. During the 1920s, the development of brine refrigeration greatly improved the ability of distant-water fleets, especially the Japanese and Americans, to fish further from home and to spend extended periods at sea. With the spatial extension of fishing effort came conflict over access to territorial waters.

As the 1930s drew to a close, the vessels of both nations were active in the Pacific. The Americans were fishing for salmon and halibut in Alaska and off British Columbia, while tuna boats from San Pedro and San Diego had moved south to work off Mexico, Peru and the Galápagos Islands. They landed catches in Honolulu and Manila. Japan, the world's leading fishing nation, despatched distant-water fleets to the waters off Korea, Taiwan, South Sakhalin and the Kurile Islands. The Japanese had long established rights to fish in Kamchatka and other Far Eastern areas of the Soviet Union (Kasahara and Burke, 1973, p14), and worked the thousand miles of coral atolls and volcanic islets of the Caroline, Marshall and Mariana Islands. Starting in 1935, they signed fishery agreements with Argentina and Mexico. Exploratory fishing was undertaken in the Bay of Bengal, and along the Australian coast, while the Japanese also had a fleet of mother ships canning king crab in the waters of Bristol Bay, which were then international despite their proximity to the west coast of Canada and the US (Borgstrom, 1964, pp30–39). In 1936, they announced that they were embarking on a three-year scientific investigation into Bristol Bay's salmon resource (Gregory and Barnes, 1939, p3), which caused political uproar in the province of British Columbia and the states of Washington and Oregon. There were already concerns about over-fishing in Bristol Bay, where fishermen were battling the politically powerful canning industry over the use of fish traps (Cooley, 1963, p157). Since 1919, biologists had warned that salmon stocks were being over-fished and urged the federal government, which controlled the fishery, to take steps to limit the number of fishermen and processors (Gilbert and O'Malley, 1919). Japan acquiesced to the State Department's request to abandon its plans to launch the investigation, and the fisheries issue was lost amid other tensions.

With the end of the war in 1945, salmon industry leaders in the Pacific Northwest urged the State Department to make sure that the Japanese would not be back. The result was the Truman Proclamation, issued in September 1945, which declared that the US had the right to create conservation zones in the high seas, where vessels from new countries would not be allowed to join established fisheries.

The proclamation was designed to keep the Japanese away from Alaskan salmon and halibut; but it also unleashed a series of territorial claims from Latin American nations, unhappy about American tuna boats fishing off their coasts, just as the Japanese had threatened to do off Bristol Bay. A month after the proclamation was issued, Chile adopted a 200-mile (320km) territorial water zone, with Peru, Mexico and Costa Rica soon following suit. In May 1949, Mexico seized gulf shrimp boats fishing south of the mouth of the Rio Grande, a clear challenge to America's cherished doctrine of the freedom of the seas: the right of merchant shipping, navies, submarines and fishing boats to move without restriction across the world's oceans.

At the same time, the US fishing industry fought for greater recognition of its problems by the State Department. As negotiations continued over a peace treaty with Japan, the salmon industry wanted to exclude Japanese boats from Bristol Bay, while the tuna industry worried that a trickle of imported tuna from Japan would soon become a flood that would sweep market share away from US processing companies. And in New England, imports of fillets from Canada, Iceland and Norway were on the increase, depressing prices for local fishermen and processors. US fishing interest saw itself in crisis and argued that fish was not just a commodity, and that fishing involved issues of international diplomacy and law. This case was accepted by Secretary of State George Marshall, who authorized the appointment of a temporary under-secretary of state for fisheries affairs. The State Department wanted a lawyer; but in the summer of 1948 it accepted a scientist, Wilbert McLeod Chapman (1910–1970). Chapman hailed from Kalama, Washington, and had received a doctorate in ichthyology from the University of Washington in Seattle in 1937. He had worked for the Washington Department of Fisheries and the California Academy of Sciences before accepting a wartime position that sent him to the equatorial Pacific for 18 months, where he scouted for fresh fish to feed American servicemen. He had recently been appointed director of the School of Fisheries at the University of Washington. On moving to Washington, DC, the most serious of the many problems confronting Chapman was the dilemma that the Truman Proclamation had created for the US fishing industry – containing Japanese fishing off Bristol Bay, but setting a precedent that was being used against American tuna boats off Latin America. Chapman's solution was the implementation in early 1949 of the US High Seas Fisheries Policy, the scientific basis for which was MSY.

Defining Maximum Sustained Yield

MSY is very much a legacy of the Progressive movement, which flourished in the US between 1890 and 1920. Progressives such as forester Gifford Pinchot and US President Theodore Roosevelt saw natural resource management as an essentially technical task that should be handled by experts guided by the ideals of efficient utilization and the greatest good for the greatest number of people

(Hays, 1959, p3). This thinking imbued the scientific establishment of the 1950s, when great faith was placed in the ability of scientists to control and manipulate nature (Rozwadowski, 2002, p146). After the devastation of World War II, the focus of scientists and policy-makers alike was on increasing the food supply with the objective of eradicating world hunger, which was seen as a cause of war. Critical to the expansion of the protein supply was increasing the harvest from the seas. During the 1950s, the global fish catch grew by 6.8 per cent a year, from 18.5 million metric tonnes (mt) in 1950 to 31.2 million mt in 1959 (Garcia and Newton, 1997). Catches on this scale were seen as conservative, sustainable and in accord with the science of MSY.

The idea of taking the maximum number of fish without hurting the reproductive capacity of the stocks had a long history in American fisheries management. The first US fish commissioner, Spencer Fullerton Baird, advocated the concept in the 1880s (Allard, 1978, p294). Its deployment was enshrined in the adoption of the US High Seas Fisheries Policy, which was announced by Chapman in a speech in San Francisco on 2 December 1948 and published in the *Department of State Bulletin* on 16 January 1949 (Chapman, 1949). On 25 January 1949, a fisheries treaty was signed with Mexico, while an agreement to regulate fishing in the North Atlantic through the International Commission for North Atlantic Fisheries (ICNAF) was reached with 11 European countries on 28 January 1949. On 31 May 1949, the Inter-American Tropical Tuna Commission (IATTC) was created by virtue of a convention signed with Costa Rica. The policy goal of all three agreements, as Chapman stated, was to manage fisheries through the attainment of MSY:

> *The policy of the United States government regarding fisheries in the high seas is to make possible the maximum production of food from the sea on a sustained basis year after year. So stated, the policy is extremely simple, and it is doubtful that there will be any objection to it from any quarters.* (Chapman, 1949)

This policy rested on a scientific formulation; but the document did not cite any scientific papers or references, even for the graph that supported the MSY theory (see Figure 11.1). Chapman was famous for the letters he wrote, always long, and often sent to multiple recipients. But there is nothing in the surviving correspondence to indicate that Chapman sought advice on the goal of US fisheries policy. The idea of harvesting the maximum amount was the logical policy choice for scientists during the Progressive and post-1945 eras, for conservation meant utilization, and the mobilization of the fisheries to contribute to the expansion of the world's food supply. Chapman also believed that fisheries science would soon have the ability to gauge when stocks were reduced and regulations to slow fishing ought to be imposed. However, the mathematical formulae to establish MSY levels had not been published in 1949, when Chapman adopted the policy.

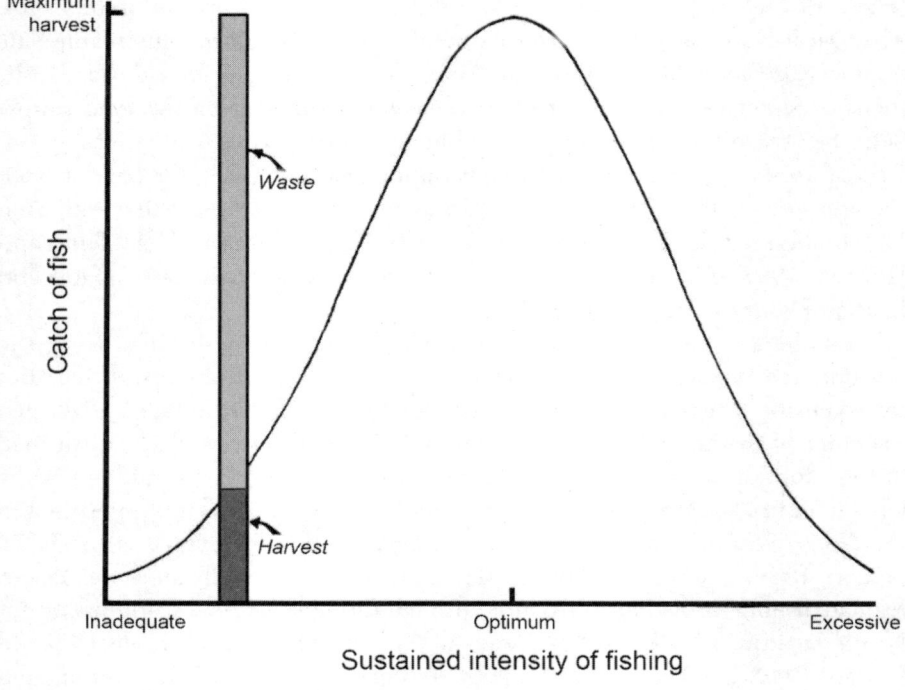

Figure 11.1 *US High Seas Fisheries Policy*

Source: Chapman (1949)

Chapman's explanation was that as the intensity of fishing on a stock increased, the reproductive capacity of the fish population also rose for reasons that scientists did not understand, perhaps because there was more food for the fish that were left or because there was less predation. He expressed the relationship in a logistics curve:

> *The meaning of this curve is that for any particular population of fish there is an optimum point of fishing intensity which, if sustained, will yield the maximum crop of fish year after year. Less fishing is wasteful, for the surplus of fish dies from natural causes without benefit to mankind; more fishing is wasteful because it depletes the population and so results actually in a smaller crop.* (Chapman, 1949, p70)

For Chapman, management was only necessary when fishing intensity had passed the point of maximum return. But not to harvest the available stocks was wasteful, as the fish died and provided no benefit to mankind. Chapman expanded on his policy in a statement in the *Fishing Gazette,* saying that it 'would be morally as well as legally unjustifiable for a resource of the high seas to be fenced off and not fished to the full extent that is needed to produce the maximum sustained harvest from the resource' (NARA, undated).

This statement was aimed at Latin American nations, who were keen to restrict US fishing. Such a threat lay behind the formulation of the multilateral agreements that underpinned both ICNAF and the IATTC. In essence, the US acted to forestall any discussion concerning the establishment of an international commission with wider powers, including, perhaps, authority to rein in America's fisheries. In particular, Chapman did not want the US to come under the authority of the UN's Food and Agricultural Organization (FAO), which the Americans had been instrumental in creating in 1945. Accordingly, the purpose of the new commissions was announced by the US before any scientific discussion could begin, even though the European scientists involved in the creation of ICNAF were primarily intent on limiting mesh sizes and fishing zones in their endeavours to deal with over-fishing issues in the Northeast Atlantic.

In 1951, Chapman left the State Department to work as director of research for the American Tuna Association (ATA), based in San Diego. He was replaced by William C. Herrington (1903–1991), who had graduated from Stanford University in 1927 with a degree in zoology, and worked for the California Fish and Game Commission and the US Fish Commission. Between 1947 and 1951, he served in occupied Japan as chief of fisheries, Division of Natural Resources, Supreme Commander Allied Powers. Herrington inherited a high-seas policy that did little to alleviate the growing fisheries tensions with Latin America.

ESCALATING TENSIONS IN GLOBAL FISHERIES

The expansion of distant-water fishing also created friction in the North Atlantic. Two developments were particularly significant in this region during the post-war years. First, in 1948, Iceland acted to claim a 200-mile (320km) zone, based on the precedent set by the Truman Proclamation in 1945. During the following year, Iceland's delegation to the UN requested that the International Law Commission study the issue of territorial claims. Despite strong opposition from the Western European nations fishing in Icelandic waters, the UN backed Iceland's motion. Icelandic scholar Hannes Jonsson considered that this decision was an important milestone in the evolution of the Law of the Sea:

> *... by adopting this resolution, the General Assembly implicitly agreed with the Icelandic delegation that the opinion of the leading West European powers – that the 3-mile limit was firmly established rule of international law of the sea – was wrong, and accepted the necessity of studying the regime of territorial waters.* (Jonsson, 1982, p56)

Accordingly, when the commission issued a draft report in the summer of 1952, it recommended that territorial limits be expanded to 6 miles (9.7km) from the current 3 (4.8km), and that an international framework should be established under the FAO to protect fish resources from waste or extermination on the high seas. Iceland began to restrict foreign fishing, primarily by the UK, within its waters.

Second, the operations of British trawlers were also at the heart of a dispute with Norway that had been simmering since these vessels first fished off the Norwegian coast in 1906. Diplomatic discussions about rights of access to these inshore waters had taken place, and in 1935 Norway had issued a unilateral decree that extended her territorial waters to a 4-mile (6.4km) belt seaward from straight baselines drawn between 48 points on Norwegian promontories, islands and rocks. The result was the enclosure of a large area of water that had formerly been regarded as the high seas and an important UK fishing ground. The UK applied to the International Court of Justice at The Hague for a ruling, and in 1951 the court upheld the Norwegian position (Green, 1952, p373). The decision was one of the first successful attempts by a country to restrict foreign fishing on territorial grounds, and it was a substantial blow, not only to the UK's distant-water fleet, but to the US's distant-water claims as well.

The actions of Iceland, Norway and the Law Commission encouraged the Latin Americans in their quest to control their own territorial waters. Tensions in the region had increased considerably in 1950, when a large new tuna ground was discovered on the Guayaquil Banks, off northern Peru. American fishing boats immediately started to pour into the area. On 6 November 1950, Ecuador declared a minimum extension of its territorial waters to 12 miles (19km) (*CFR*, 1952), a measure that was reinforced by the promulgation of a further decree on 29 January 1952 that prohibited any foreign fishing vessels from entering, for any purpose, a 12-mile zone along the Ecuadorean coast. The zone included the Galápagos Islands, an important tuna fishing ground for American boats.

The looming threat to US fisheries intensified in August 1952 at a meeting of the South Pacific Fishery Conference in Santiago. Here, delegates from Chile, Ecuador and Peru (with Colombia sending an observer) acted decisively to assert wider territorial limits. The group also voted to establish a permanent commission to superintend fishery operations in the territorial waters of the participating nations. The new commission would have authority to undertake technical investigations and make recommendations to governments, to restrict fishing and to establish quotas, prohibited areas and closed seasons (NARA, 1952a). Although these countries were acting to exclude US, Japanese and European vessels, Chile's chief concern was protection from a fleet of Soviet whaling ships that was operating outside the territorial boundaries set by the International Whaling Commission, while Peru and Ecuador were more concerned about US tuna boats in their waters (NARA, 1952b).

While the initial thrust of the new commission was to regulate whaling, the Santiago Declaration expounded the concept of territorial waters. This was taken a stage further at the Tenth Inter-American Conference at Caracas in 1954, when Venezuela proposed a motion to establish a 200-mile (320km) limit for the territorial sea in Latin America. The Americans managed to keep the proposal from a vote. But the meeting decided to ask the Organization of American States to hold an International Conference on the Living Resources of the Sea in 1955 (Hollick, 1981, p88). The US was alarmed at these developments. While it perceived unilateral action to establish expanded territorial limits in the sea as jeopardizing its high-seas tuna fishery, and providing encouragement for other countries to restrict American actions, the prospect of a conference raised the spectre of the establishment of an international organization with binding authority.

The US response was to try and take control of the conference by moving it into the jurisdiction of the UN, by changing its location to Europe to dilute the Latin American presence, and by dictating the agenda. The vehicle for shifting the conference was the 1953 report by the International Law Commission. Given the escalating number of territorial conflicts, the commission stated that new international legislation was needed; but since the 'technical character' of fisheries was beyond its expertise, it recommended that an international conference be convened to provide the necessary technical advice. Chapman and Herrington began enlisting support for a meeting that would answer the commission's call for action without leading to the creation of a supra-national fisheries management body. Herrington advocated the notion of an international fisheries conference to the UN General Assembly in December 1954. He proposed that it be held in Rome, at FAO headquarters, during the spring of 1955. Timing was of the essence for Herrington wanted the conference to conclude by early May so that its recommendations could be forwarded to the International Law Commission before that body met in Geneva in June (NARA, 1955a). The commission would then make recommendations to the UN 11th session in 1956.

Such a schedule would pre-empt and circumvent the conference planned by the Organization of American States. Moreover, it left very little time for delegates to prepare for a conference that would take place against a background of complex legal, political and diplomatic issues. While the illegal whaling fleet of Aristotle Onassis had been arrested by the Peruvian navy (Fraser et al, 1977, p123), the Soviet Union had claimed a 12-mile (19km) territorial sea (which the US refused to recognize), the Australians suspected that the Japanese were fishing in their waters, the Russians and the Japanese were contesting access to the waters off Sahkalin, and the Japanese were in dispute with Korea over territorial limits. In this context, the Americans wanted to keep the agenda as narrow as possible so that delegates did not get bogged down in regional disputes or in the question of territorial claims (FAO, 1954). This was reflected in the UN General Assembly's terms of reference, which stipulated that the conference should deal with the technicalities of fishing, rather than legal or political matters (FAO, 1955a, piii). Scientists were therefore invited to write papers on various issues relating to conservation and how it was being implemented.

Herrington headed the American delegation, which included Chapman. Before the meeting he travelled to Ottawa, London, Havana and several Scandinavian countries to make potential allies aware of the American position. This was explicit in the instructions issued to US delegates: 'it will be the task for the US Delegation to bolster the declining case for the 3-mile limit by the forceful and logical presentation of the fishery conservation principles for which the US stands' (NARA, 1955b). Moreover, Herrington made it clear that the US was opposed to the appointment of an international organization with binding powers. The Americans advocated regional fishery agreements, such as the International Halibut Commission set up by the US and Canada, as a better way of resolving fishing disputes. Conservation could only be achieved through scientific programmes, set up by international agreements among the concerned states. Sovereignty beyond 3 miles (4.8km) was not needed. Most importantly, the Americans wanted to separate conservation issues from any question involving territorial limits (NARA, 1955c).

By 1955, ten Central and South American countries had declared some sort of expanded jurisdiction over their coastal waters; while Mexico had asserted rights over a 9-mile (14.5km) zone, Argentina and Panama had claimed their respective continental shelves, and Chile, Peru, Costa Rica, El Salvador, Honduras, Ecuador and the Dominican Republic had declared 100-mile (161km) limits. The US had objected to these claims and reserved its right of free passage on the high seas. This entailed more than just the right to fish: national security, naval, maritime and air transport could all be affected by fishing restrictions. As America's delegates to the Rome Conference were told:

> ... it is incumbent on the United States to take every opportunity to maintain the principle that international law does not require a state to recognize more than 3 miles of territorial waters. Beyond that lie the high seas, to which freedom of navigation and fishing for all countries appertains. (NARA, 1955c)

Three weeks before the conference started, an Ecuadorian patrol ship seized two American fishing vessels between 14 (22.5km) and 25 miles (40km) west of an island off its country's coast. During the seizure, an American seaman was seriously wounded by gunfire from the patrol ship. Fines of more than US$49,000 were imposed on the fishing vessels despite strong American protests (*CFR*, 1955). Tensions were high as delegates headed to Rome.

The Rome Conference

The conference commenced as scheduled on 18 April 1955. Delegates from 45 countries attended, including the Soviet Union, as well as scientists from regional and international organizations such as the International Council for the Exploration of the Seas. There were Cold War undertones. The Soviet Union proposed that delegates from Communist China, East Germany and North Korea should attend. The Americans opposed, saying only UN members could participate. There was also conflict with Latin American states. On 2 May 1955, a telegram to the State Department stated:

> *Extreme position three South American states on fisheries controls is manifesting itself this stage and being supported by Soviet bloc in obvious attempt to promote recognition special rights coastal states and in general block satisfactory recommendations to ILC by conference.* (NARA, 1955d)

The delegates were almost evenly divided between two groups; on the one hand, there were coastal countries, such as Peru and Mexico, which did not have developed fisheries, while on the other there were the major fishing nations, including the US, Japan, the UK and Norway, which had very effective, wide-ranging fisheries. Scientific differences were also evident. Many of the Europeans, concerned about over-fishing in the North Sea, supported conservation by managing effort. The Americans focused on the work of scientists such as Milner B. Schaefer, director of the IATTC and one of the most respected scientists of his day. Schaefer had recently published his 'surplus production' theory, through which the maximum total harvest of fish every year from a standing population could be estimated (Smith, 1994, p3). The theory also postulated that all fish populations had fish that are 'surplus' to the spawning biomass and can safely be harvested. Schaefer believed that fish populations tended to stay in balance with their environments. If more fish were taken by predation or other means, the population would tend to come back into balance. Schaefer also saw fishing by humans as part of the natural predation that fish faced. He argued:

> *It is simply an increase in the predation rate. It is the compensatory reaction of the fish population to the mortality produced by fishing which made a fishery possible, so that the population comes into balance under environmental conditions which include predation by man. It is theoretically possible to impose so much fishing ... to drive it below its threshold magnitude for survival. I know, however, of no instance where this has been accomplished in a purely marine fishery. It appears that the threshold magnitude is almost always well below the population size to which it is economically possible to fish.* (Schaefer, 1955, p15)

Schaefer was following the basic argument laid out by Wilbert Chapman in the 1949 US High Seas Fisheries Policy, which had enshrined MSY as the goal of American fisheries policy. Schaefer equated fishing by humans as a natural part of predation in the ocean, in keeping with MSY's inherent implication that fishing is good for fish stocks, creating the conditions that produce large numbers of younger fish that grow rapidly to a harvestable size. Schaefer recognized that the size of fish populations was affected by other variables, including environmental factors, and therefore one of the chief tasks for scientists was to account for natural fluctuations in stocks, and isolate them from the impact of fishing.

For the US, the most significant debate revolved around three paragraphs that would frame the conference's recommendations to the International Law Commission. The language was telling:

1. *Conservation is essential in the development of a rational exploitation of the living resources of the seas. Consequently, conservation measures should be applied when scientific evidence shows that fishing activity adversely affects the magnitude and composition of the resources or that such effects are likely.*
2. *The immediate aim of conservation of living marine resources is to conduct fishing activities so as to increase, or at least maintain, the average sustainable yield of products in desirable form. At the same time, wherever possible, scientifically sound positive measures should be taken to improve the resource.*
3. *The principle objective of conservation of the living resources of the sea is to obtain the optimum sustainable yield so as to secure a maximum supply of food and other marine products.* (NARA, 1955e)

By a one-vote margin, the conference adopted a proposal submitted by Mexico and Peru to amend the first paragraph by the insertion of the following sentence:

When formulating conservation programmes, account should be taken of the special interests of the coastal state in maintaining the productivity of the resources of the high seas nears to its coast. (FAO, 1955b)

On 5 May, the motion passed by 18 votes to 17, with 8 abstentions. Chapman's notes infer the divisive and confrontational character of the debate: 'against us', he recorded, were Argentina, Brazil, Chile, China, Colombia, Costa Rica, Cuba, Ecuador, Guatemala, Iceland, India, Indonesia, Korea, Mexico, Paraguay, Peru, Uruguay and Yugoslavia; 'for us', were Belgium, Egypt, France, Germany, Greece, Italy, Japan, Monaco, The Netherlands, Norway, Poland, Portugal, Spain, Sweden, the UK, the USSR and the US. The abstaining nations were Australia, Canada,

Denmark, Honduras, Nicaragua, Panama, South Africa and Turkey, with El Salvador and Israel absent (SIOA, undated).

The thrust of the recommendations reflects the scientific thinking at the time: that fishing and other human activities, such as hatcheries, played a positive role in increasing or maintaining fish stocks. The resource could be improved to increase its utility to humankind. Science would be able to estimate when fishery resources merited protection. How protection would be applied was not discussed. In a memo written after the conference, D. B. Finn, the FAO's fisheries director, highlighted the significance of the first paragraph. Conservation measures would not be justified without scientific evidence. The conference had decided that the aim of fisheries management should be to harvest fish until a critical maximum point was reached, when conservation measures could be applied. Accordingly, MSY reflects the underlying belief that fish stocks are robust and resilient, and that not to harvest is wasteful. Fishing is viewed as a benign force in ecosystems, removing older fish that compete with younger cohorts for space and food (FAO, 1955b).

When the Rome Conference ended, the leaders of the American delegation, William Herrington and Wilbert Chapman, drove to Geneva to attend the International Law Commission meeting. The commission accepted the 'technical' advice of the Rome Conference, and MSY became a legal concept, as well as a policy objective and scientific principle. This multifaceted character partly explains the continuing sway of MSY. A theoretical concept that has been adopted by scientific, legal and political establishments, MSY has endured despite its inherent scientific weakness. The criticism that was levelled at MSY by delegates to the Rome Conference, and others, has been lost to sight. Moreover, since 1955, there have been no attempts to create an international body with binding authority to adjudicate in fisheries disputes.

OPPOSITION AMONG SCIENTISTS

The UK's Michael Graham, in his keynote address to the Rome Conference, examined the history of fisheries conservation and took the opportunity to repeat the central theme of his book *The Fish Gate* – that fisheries needed to be regulated or they would become unprofitable (Graham, 1943, p151). Citing the work of W. F. Thompson at the International Pacific Halibut Commission, Graham pointed out that the commission's management had resulted in an increased catch. However, there were so many fishermen that the season had shrunk from eight and a half weeks in 1932 to only 28 days on one fishing ground and 56 days on a second, and the stocks were facing fishing pressure. Fishing was obviously less profitable for the fleet.

Graham also contributed a four-page paper that contained a pointed reference to Schaefer's argument that fishing was part of the natural predation rate in the ocean. Graham opined that: 'if one agent of death becomes so active as to claim more fish than die by all other agencies together, then that agent has control of the average age of the stock of fish' (Graham, 1955, p57). He argued that when fish were caught at a later size, the yield was much higher: 'great benefits of several kinds are therefore obtainable if the rate of fishing can be controlled, especially when, as often is the case, an intermediate rate of fishing gives the best result' (Graham, 1955, p57). According to Graham, gear restrictions and mesh sizes should be introduced to the fishery early on, before scientists had established that stocks were declining, while the research was ongoing. It was a 'go slow' approach that sought to achieve long-term economic benefits for fishermen by protecting young fish from exploitation until they were older and had spawned.

In a sense, the outcome of the conference had been decided before it started, and Graham said as much in his keynote address:

> *In the Old World we have not as yet made any explicit choice among the possible qualities of a fishery – but in the New World, the choice of maximum sustained yield has been explicit in all recent international conventions.* (FAO, 1955a, p11)

In his analysis of the Rome Conference, Tim Smith argued that in 1954 three theories which contributed to the formulation of MSY were published. Each sought to estimate the impact of fishing on stocks. Most influential was the work of British biologists Raymond Beverton and Sidney Holt, whose 'yield per recruit' theory focused on how to calculate the maximum yield from each cohort of a fish population. It would prove to be one of the founding documents for the understanding of fish populations (Rozwadowski, 2002, p159). Canadian William Ricker published a 'spawner per recruit' theory that estimated the maximum number of spawners from each year class of salmon. Schaefer's 'surplus production' theory emerged from his work on sardines (for the US Fish and Wildlife Service) and tuna, and offered a means by which the maximum total harvest of fish could be estimated from a standing population (Smith, 1994). Each theory describes the life of a year class of fish in terms of logistic curves that trace the size and growth rate of the population and when natural mortality starts to reduce it. The shape of the curves depends upon how quickly the fish grow to maximum size. The objective of MSY is to harvest the maximum number of fish at their maximum weight so that no fish are wasted and the spawning biomass is protected. Smith argued:

> *Schaefer's research paradigm formalized a narrow approach that fishery research has been slow to change from. This narrow focus on populations of fish in isolation was reflected in the terms of reference of the many international agreements for the study and management*

of marine fisheries that were put in place in the late 1940s and early 1950s. (Smith, 1994, p335)

This analysis requires modification. The main outcome of the Rome Conference, especially for the non-scientists who controlled the national votes, was that stocks had a harvestable surplus that scientists could estimate accurately, which meant that fishing activity could be regulated when the need arose, not before. Schaefer's formulation solved one of the most perplexing problems that had faced managers: how to determine a fishery's maximum sustained yield, or MSY. His mathematical framework explained the relationship between fishing effort and catch, and allowed scientists to estimate the proportion of the biomass that could be removed by a single fisherman, while taking into account the stock's ability to increase and its maximum theoretical size. These estimates could then be used to predict both the maximum average yield that the stock could attain for any level of fishing effort (McGoodwin, 1991, p69).

But Smith is correct in concluding that the partial theories that underpinned MSY could not be melded into a successful research programme. However, this is because MSY is not based on science at all. It is an inherently political construct designed to achieve a specific objective – that is, allowing the distant-water fleets of developed countries to continue to take fish from the coastal waters of undeveloped, poorer countries. As such, it has been spectacularly successful in permitting two developed countries, Japan and the US – the leading tuna fishing nations – to continue to dominate the catch of high-seas tuna for the next several decades, even though it has been much less effective in other contexts.

Schaefer's theoretical work should have been tested by scientists before it was incorporated into policy. But it was not, even though scientists at the time knew that some fish populations did exhibit large dome-shaped curves. In fact, sometimes the curves were flat or asymptotic, indicating that very little or none of the population was surplus to reproductive needs. Nevertheless, 'surplus production' theory was embedded in the Rome Conference's conclusions regarding the process of fisheries management. There would not be an international organization, with binding authority, to mediate fishery disputes. Conservation would come through limited, bilateral or multilateral agreements relating to specific stocks. While some fish would be protected, vast areas of the oceans would not (Gjerde and Freestone, 2004). Fish would be managed for MSY, which now included an assumption that a percentage of each population could safely be harvested.

Conclusion

Evidence from a range of archival sources indicates that the US actively manoeuvred to have MSY adopted as the goal of international fisheries management at the Rome Conference. The freedom of the seas, and the right to fish in international waters, was a hotly contested issue during the 1950s, as countries with underdeveloped fisheries sought to limit predation off their shores by the distant-waters fleets of developed nations – notably, the US, Japan and the UK. The relevant State Department documents invariably concern the policy (rather than the scientific) implications of MSY, and imply that the Americans used fisheries science as a tool to bolster the policy of the freedom of the seas and to prevent Latin American countries from regulating the fishing activities of US vessels (Hannesson, 2004, p40). Unilateral action taken by any country which threatened to impede such activities was deemed undesirable.

The Rome Conference not only served to place the burden of conservation solely on a science that was in its infancy, but also firmly directed that science towards the quantification of marine animals, rather than the enhancement of knowledge and understanding of their role in ecosystem dynamics. But most of all, the Rome Conference established a pattern of fisheries management that has persisted since 1955; in essence, managers and politicians have not imposed any constraints on fishing effort until critical biological points have been reached. Scientists have been too optimistic about their ability to identify these critical points in naturally fluctuating systems. Accordingly, the managerial regime founded on MSY has contributed to an enormous over-capacity in world fisheries, economic waste and depleted stocks.

REFERENCES

Allard, D. C. Jr. (1978) *Spencer Fullerton Baird and the US Fish Commission*, Arno Press, New York
Berkeley, S. A., Hixon, M. A., Larson, R. J. and Love, M. S. (2004) 'Fisheries sustainability via protection of age structure and spatial distribution of fish population', *Fisheries*, vol 29, no 8, pp23–32
Borgstrom, G. (1964) *Japan's World Success in Fishing*, Fishing News (Books), London
CFR (1952) *Commercial Fisheries Review*, October, p61
CFR (1955) *Commercial Fisheries Review*, September, p5
Chapman, W. M. (1949) 'United States policy on high seas fishing', *Department of State Bulletin*, vol XX, no 498, 16 January, pp67–80
Cooley, R. A. (1963) *Politics and Conservation: The Decline of the Alaska Salmon*, The Conservation Foundation, New York
Doel, R. E. (1997) 'Scientists as policymakers, advisors, and intelligence agents: Linking contemporary diplomatic history with the history of contemporary science', in T. Soderqvist (ed) *The Historiography of Contemporary Science and Technology*, Harwood Academic Publishers, Amsterdam, pp215–244
FAO (1954) RG 14, 4 October, Food and Agricultural Organization of the United Nations, Rome
FAO (1955a) *Papers Presented at the International Technical Conference on the Conservation of the Living Resources of the Sea*, 18 April–10 May, Food and Agricultural Organization of the United Nations, Rome

FAO (1955b) O-163 RG 61.1, Series C3, memorandum from D. B. Finn to Director-General of FAO, 19 May, Food and Agricultural Organization of the United Nations, Rome

Fraser, N., Jacobson, P., Ottaway, M. and Chester, L. (1977) *Aristotle Onassis*, Times Newspapers, London

Garcia, S. M. and Newton, C. (1997) 'Current situation, trends, and prospects in world capture fisheries', in E. K. Pikitch, Daniel D. Huppert and M. P. Sissenwine (eds) *Global Trends, Fisheries Management*, American Fisheries Society, Bethesda, MA, pp3–27

Gilbert, C. H. and O'Malley, H. (1919) 'Special investigations of salmon fishery in central and western Alaska', *Alaska Fisheries and Fur Industries*, pp143–160

Gjerde, K. M. and Freestone, D. (2004) 'Unfinished business: Deep-sea fisheries and the conservation of marine biodiversity beyond national jurisdictions', *International Journal of Marine and Coastal Law*, vol 19, no 3, pp209–222

Graham, M. (1943) *The Fish Gate*, Faber and Faber, London

Graham, M. (1955) 'A first approximation to a modern theory of fishing', in FAO (ed) *Papers Presented at the International Technical Conference for the Conservation of the Living Resources of the Sea*, Rome, pp56–58

Green, L. C. (1952) 'The Anglo-Norwegian fisheries case, 1951', *Modern Law Review*, vol 15, no 3, pp373–377

Gregory, H. E. and Barnes, K. (1939) *North Pacific Fisheries*, American Council, Institute of Pacific Relations, San Francisco and New York

Hannesson, R. (2004) *The Privatization of the Oceans*, MIT Press, Cambridge, MA

Hays, S. P. (1959) *Conservation and the Gospel of Efficiency*, Harvard University Press, Cambridge, MA

Hollick, A. L. (1981) *US Foreign Policy and the Law of the Sea*, Princeton University Press, Princeton, NJ

Jonsson, H. (1982) *Friends in Conflict: The Anglo-Icelandic Cod Wars and the Law of the Sea*, C. Hurst & Co, London

Kasahara, H. and Burke, W. (1973) *North Pacific Fisheries Management: The Program of International Studies of Fishery Arrangements*, Resources for the Future Inc, Washington, DC

Krige, J. (2006) *American Hegemony and the Postwar Re-construction of Science in Europe*. MIT Press, Cambridge, MA

Larkin, P. A. (1997) 'An epitaph for the concept of maximum sustained yield', *Transactions of the American Fisheries Society*, vol 106, no 1, pp1–11

McGoodwin, J. R. (1991) *Crisis in the World's Fisheries: People, Problems, and Policies*, Stanford University Press, Stanford

Myers, R. A. and Worm, B. (2003) 'Rapid worldwide depletion of predatory fish communities', *Nature*, vol 423, pp280–283

NARA (undated) RG59 811.245/1-950, Box 4425, US National Archives and Records Administration, College Park, MD

NARA (1952a) RG 59, folder 398.245-SA/8-2052, Report from Santiago Embassy to State Department, 20 August, US National Archives and Records Administration, College Park, MD

NARA (1952b) RG 59, 398.245 SA/9-552, Foreign Service Despatch, Santiago Embassy to State Department, 5 September, US National Archives and Records Administration, College Park, MD

NARA (1955a) RG 59, Box 1538, Central Decimal Files, 1955–1959, Memorandum, 31 January, US National Archives and Records Administration, College Park, MD

NARA (1955b) RG 59, Box 1538, Central Decimal Files, 1955–1959, Memorandum, 11 March, US National Archives and Records Administration, College Park, MD

NARA (1955c) RG 59, Box 1539, folder 398.245-SA6-755, State Department Memorandum, William Herrington to Harold Kissick, 7 June, US National Archives and Records Administration, College Park, MD

NARA (1955d) RG 59, Box 1538, folder 398.245-RO/5-255, Embassy telegram, 2 May, US National Archives and Records Administration, College Park, MD

NARA (1955e) State Department Memorandum from William Herrington, 30 November, US National Archives and Records Administration, College Park, MD

Pauly, D. (1994) *On the Sex of Fish and the Gender of Scientists: A Collection of Essays in Fisheries Science*, Chapman and Hall, London

Pauly, D. and Maclean, J. (2003) *In a Perfect Ocean: The State of Fisheries and Ecosystems in the North Atlantic Ocean*, Island Press, Washington, DC

Rozwadowski, H. M. (2002) *The Sea Knows No Boundaries: A Century of Marine Science under ICES*, University of Washington Press, Seattle, WA

Schaefer, M. B. (1955) 'The scientific basis for a conservation program', in FAO (ed) *Papers Presented at the International Technical Conference for the Conservation of the Living Resources of the Sea*, Rome, pp15–55

SIOA (Scripps Institution of Oceanography Archives) (undated) American Tuna Association Files, Box 76, 'Rome' folder, University of California, San Diego, CA

Smith, T. (1994) *Scaling Fisheries: The Science of Measuring the Effects of Fishing, 1855–1955*, Cambridge University Press, Cambridge, UK

Afterword

Lost and Found in the Past

Paul E. Waggoner

History, as this volume demonstrates, offers critical insights to those who seek to understand the current and future state of life in the oceans. The past also holds lessons for contemporary census-makers, especially scientists intent on compiling a Census of Marine Life (CoML). The highly focused objectives of early censuses generated relatively narrow results. Ancient Babylon, Palestine, Persia, China and Egypt restricted surveys to particular categories, such as family heads or males of military age. The Roman census of 2 to 3 BC was both a counting of men and an oath of allegiance:

> *[Augustus] ordered that a census be taken of each province everywhere and that all men be enrolled.... This is the earliest and most famous public acknowledgement which marked Caesar as the first of all men and the Romans as lords of the world.* (Pratt, 1990)

Such a headcount of men, or such an oath, might help to raise an army or command the loyalty of an empire. Yet, because of its narrow parameters, the census reveals only limited information on the structure and trends of a society.

In England, the *Domesday Book* was exceptional in scale and scope. Twenty years after William of Normandy invaded England in 1066, this exercise sought to inventory his conquest:

> *Here is subscribed the inquisition of lands as the barons of the King have made inquiry into them; that is to say by the oath of the sheriff of the shire, and of all the barons and their Frenchmen, and the whole hundred, the priests, reeves, and six villeins of each manor; then, what the manor is called, who held it in the time of king Edward, who holds now; how many hides, how many plows in demesne, how many*

> *belonging to the men, how many villeins, how many cottars, how many serfs, how many free-men, how many sokemen, how much woods, how much meadow, how many pastures, how many mills, how many fish-ponds, how much has been added or taken away, how much it was worth altogether at that time, and how much now, how much each free man or sokeman had or has.* (Halsall, 1996)

For CoML, the example of the *Domesday Book* teaches that the abundance and distribution of one species does not suffice, for William's inquisitors counted priests and villains, ploughs and fish ponds, meadows and woods, and then reckoned how much it was all worth. In essence, *Domesday* shows how a broad census has the potential to reveal the structure of an ecosystem, while its analysis can elucidate that system's functioning. It challenges marine scientists to document and then decipher how other species and the environment affect the landings of one taxed species or the salt to preserve it. More specifically, it challenges the History of Marine Animal Populations project (HMAP) – the historical component of CoML – to calculate 'how much has been added or taken away, how much it was worth altogether at that time, and how much now': in other words, to measure change over time.

LIMITS TO KNOWLEDGE

For the census, *knowledge* is placing a form of life in an environment and on the evolutionary tree with its physiology understood. Knowledge is observing a collection of life forms that live together, and grasping interactions within the collection and thus predicting its communal behaviour with a reasonable probability. But what are the *limits* to knowledge? Pondering this question can guide us towards doors to the unknown, rather than leaving us pushing at doors that are already ajar. It can save us the effort and frustration entailed in kicking at those indestructible doors that enclose the unknowable. Knowing the nature of limits can facilitate the opening or unlocking of the doors. As Jesse H. Ausubel outlines in the Foreword to this volume, a taxonomy of five families illuminates the genera of limits to knowledge. Elaborating upon the characteristics of these taxonomic groups, with emphasis placed on the challenges they present to census-makers and historians alike, is a fitting way to reflect upon approaches, insights and prospects offered by this volume.

Vastness of the present state

The vastness of the present state begins with the hundred million million million-fold ranges of size and speed. A 100 tonne whale speeds at 50km per hour amid microscopic bacteria that only drift with the current. Such spectra divert attention from the small and slow to the large and swift (O'Dor, 2003, p8). Demanding diverse skills and equipment, the ranges of mass and speed impair effective census-making with regard to abundance, whether measured as populations or mass.

The instructions for the *Domesday Book* specified: 'the priests, reeves, and six villeins of each manor; then, what the manor is called, who held it in the time of king Edward' (Halsall, 1996). The handles of address and time are crucial for a census. Interconnections among seas, plus their interconnections with rivers and lakes, amplified by currents and migrations, however, vitiate the handles of address and time for a marine census. The interconnections can separate cause from effect, and confuse assignment of a cause and invalidate an experimental treatment.

The 'butterfly effect' of sensitive dependence upon initial conditions is the essential quality of chaos. It reveals how imprecise initial conditions limit knowledge (Hilborn, 1994). Unfortunately, the volume and expanse of the oceans limit a complete exploration. The vastness of the marine environment presents the special danger of extrapolation to depths, latitudes and longitudes beyond one's observations. The darkness, and especially the pressurized depths, of oceans limit exploration by air-breathing mankind, putting a premium on technology for any census of marine life.

While many of CoML's field projects are exploiting the potential of technological developments (www.coml.org), *Alvin*, an example from the recent past, shows how technology can ignite revolutions in marine science. Delivered to the Woods Hole Oceanographic Institution in 1964, *Alvin* was a 25-foot (8m) submarine that could dive deeper, was more mobile and had more windows than her predecessors. When her original aluminium frame was replaced by titanium, she could dive to 4500m. In 1977, *Alvin* descended 2700m to the Mid-Atlantic Ridge, where the crew saw sights no human had ever seen. Large white clams and white crabs were living in hydrogen sulphide-saturated water. At the site that they called 'The Garden of Eden', they saw an animal that looked like a dandelion gone to seed. They saw worms three metres long, with bright red plumes emerging from tube-like white shells. All were thriving in pitch darkness without photosynthesis and alongside vents spewing plumes of 350 °C water. And all were visible to people because technology allowed them to penetrate the vastness of the ocean realm (Ellis, 2005).

Reasoning from the parts to the whole

Errors in measuring components propagate larger errors when they are combined in a whole system. Imagine that you have composed a marine model from its intricate parts. Mysteriously, however, the natural system may stubbornly change more violently or more abruptly than your model composed of the system's parts – despite your best efforts to specify its parts precisely. Unable to explain the abrupt changes, you may evoke emergent properties or extreme system sensitivity.

You may say that the system has lost its resilience to tolerate disturbance without collapsing into a different state. Or you may attribute the abrupt changes to a sensitive dependence upon initial conditions and chaos. Predicting, even analysing, the statistics of such extremes as 100-year floods (which are, by definition, rare) is notoriously difficult. Finally, you may frankly say that the inability to scale up predictions from the parts to the whole limits knowledge.

In broad terms, hysteresis – or path dependence – is the lagging of an effect behind its cause. It manifests itself when a result depends upon the history of a process. Different paths of onset and recovery limit knowledge of marine changes and recoveries. For example, at the same level of toxic pollution, marine life may be more abundant when water quality is worsening than when it is improving. Or, at the same level of fishing effort, the population of fish may be more abundant when fishing effort is increasing than when it is decreasing. New species inevitably invade to become components of systems, and old species depart. Pathogens notoriously invade. We know components will inevitably come and go. But we do not know the name of the invader or defector that will confound our model of a system. Alternatively, an unforeseen change in susceptibility, perhaps exacerbated by environmental change, can make a killer of a formerly harmless member of the system, confounding a model.

Extreme events not only have impacts exceeding the years of average events. They also bring into play processes that are different from contemporary ones, thereby invalidating models constructed of the system's parts and tested with contemporary observations. Downpours burst dams, plug bridges and open channels that ordinary rains cannot. The characteristic infrequency of extreme events renders them inordinately difficult to study, and waiting for them often exceeds the human span.

Surprise interventions from outside

The unpredictability of humanity can serve to render the future unknowable. Price, style and regulation all influence consumers. When consumers switch from caviar to catfish, crowd to shores rather than remain inland, or dump waste at sea rather than bury it underground, they render physical models moot. What mathematical model would have predicted that, after a moratorium on whaling, the Japanese would harvest whales under a scientific permit with the meat somehow ending in the market? What model would have predicted that the slimehead would be renamed orange roughy to serve as a replacement for declining flounder, that after 1979 it would be caught at depths of some 1000m and be 80 per cent depleted by the time New Zealand restricted its fishery? (Ellis, 2005).

Populations of millions of humans, multiplied by affluence per person, provide the economic leverage to affect marine life, accidentally or intentionally, harmfully or beneficially. Despite efforts to improve demographic and economic forecasts, their persistent inaccuracy still limits prediction of marine life. Failing technology heightens the impact of a consumer's choice; but successful technology tempers it. The unpredictability of mankind's inventions, especially of breakthrough technology, and their wide adoption, limits knowledge of their future impact on marine life (Waggoner and Ausubel, 2002).

Blinders we put on ourselves

Like fish swimming in schools, scientists congregate in disciplines where they specialize and are encouraged by fellows. Their powerful specialization inevitably carries them farther from outsiders, a disadvantage that counters some of the advantage of their specialization. In marine science, separation from inventors would limit science's use of sound waves that scan the bottom of the sea, and separation from seafarers would limit its use of vessels that could tow a sampler free of charge. Separation from historians would neglect 18th-century bills of lading, 19th-century logbooks and the diaries of 20th-century trawler-men silently awaiting discovery in archives.

Impatient with teeming, diverse marine life, scientists can oversimplify, as by replacing a physical parameter with a statistical average. Or they can deprive themselves of simplification completely, forgetting that:

> ... *the utmost effort of human reason is to reduce the principles, productive of natural phenomena, to a greater simplicity, and to resolve the many particular effects into a few general causes, by ... reasonings from analogy, experience, and observation.* (Hume, 1748)

Only self-discipline can fulfil a contract. Unfortunately, the same self-discipline could blind the contractor to serendipitous discoveries of things unspecified in the contract, thereby limiting knowledge.

Invisibility of the lost past

One easily understands that the passage of centuries removes the possibility of an instrumental record before about 1800. Less easily understood is being adrift in overwhelming seas of data. While scientists live to enlarge knowledge, this process can develop to the point where it is practically inaccessible to a curious worldwide population. Short attention spans of individuals and institutions may limit the accumulation of observations as the present promptly becomes the past. If present generations do not keep the faith and maintain databases begun by older generations, the series of data will expire, just when they are becoming valuable.

The decline or extinction of some species and the rise of others makes survivors and current populations unrepresentative of life in the past. Fossils of mussels and clams fill the geologist's bucket, while jellyfish and bacterium rot and are therefore unrecorded. A species becomes valuable and enters our ken. The fossilized species and the taxed harvest persist so that the calcified and valuable species dominate the historian's view.

Fossils represent an echo of the past, and analysis can sometimes produce surprising outcomes. The bony lobe-finned coelacanths appeared about 350 million years ago and were found as fossils in rocks from the end of the Permian period, 245 million years ago, to the end of the Jurassic, 144 million years ago. Since scientists thought that coelacanths became extinct about 60 million years ago, there was great surprise when one was caught in the Indian Ocean in 1938. The natives of the Comoro Islands, who had long been catching coelacanths, deflated their surprise somewhat. After the scientists recovered, they were able to study all the soft parts of an animal that cohabited the Earth with dinosaurs. They could study their blood, their liver, how they breathed, how they reproduced, and how they fed and swam some 65 million years ago (Ellis, 2005). Other animals in existence today may also have lived with dinosaurs; but if their ancestors decayed rather than fossilized, their prehistoric pedigree is lost in the past.

Found in the Past

Seeking to account for the ongoing degradation of the marine environment, the scientific community continues to rail against profit-seeking fishers, deaf politicians and unresponsive regulators. Yet, practitioners, policy-makers and managers should not bear all the blame. Science must also accept some culpability in advocating what has become the conventional wisdom – that is, 'stop fishing and all will be well'. Where is the irrefutable scientific knowledge that relinquishing fishing will restore Nature as she was? Without that irrefutable knowledge, can one blame fishermen for going to sea rather than staying ashore while a bank reclaims their vessels; or governments for not embargoing voters in port where idleness impoverishes them; or managers for failing to impose effective regulations on a volatile, contested and largely unknown realm?

The CoML field projects, the Future of Marine Animal Populations (FMAP) models and Ocean Biogeographic Information System (OBIS) data systems will help to correct this situation. But they will be hard pressed to produce the required irrefutable knowledge before census day in 2010. It was therefore wise of the census-makers to create HMAP in order to instil within CoML a historical dimension that promises to reduce the limits to knowledge. History embraces and informs all human life. It is therefore a multidisciplinary field of enquiry by its very character – a simple truth that goes some way towards lifting the 'blinders we put on ourselves'. When history meets science, the blinders will be lifted still further. History, moreover, can surmount the hysteresis, or time lag, between cause and effect that might serve to obscure the results of a ban on fishing or pollution and therefore discredit such a regulatory device. In this regard, the long temporal perspective provided by HMAP is instructive, for the canny politician or regulator unimpressed by colour-emblazoned charts and mind-numbing models will probably respond to robust historical evidence that hysteresis delays a good or a disappointing result of an environmental regulation.

Most significantly, history raises the cloak of 'invisibility' that envelops the 'lost past'. As this volume indicates, it finds this past in three distinct, yet interrelated, respects. First, history recovers a multiplicity of source materials, ranging from ephemeral and everyday restaurant menus (Chapter 4), to the surveys of pioneering scientists (Chapter 10), to the memoranda of hard-headed, manipulative policy-makers (Chapter 11). The second stage entails the processing of archival material to yield reliable, accessible evidence that can sustain rigorous analyses (especially Chapters 6, 7 and 9). Third, by analysing and contextualizing this novel and validated evidence, history adds depth, breadth and insight to our knowledge and understanding of the long-term interactions between human societies and marine environments.

Finding the lost past serves to warn us that, having forgotten what existed, we may simply think something different never existed. Many will be shocked and disbelieving at changes as large and diverse as those revealed in this volume: in the North Pacific, flourishing populations of right whales were destroyed within the course of the 1840s (Chapter 8); in the inshore waters of Florida, the introduction of new harvesting techniques virtually eradicated once-extensive sponge banks (Chapter 3); in the intertidal zone of northeast America, a human-mediated invasion of common periwinkle snails altered the marine ecosystem (Chapter 2); and on the banks of Cape Cod's rivers, local communities allowed the once-cherished alewife to be fished to virtual extinction (Chapter 5).

But in locating, collating and analysing information that natural science might ignore or dismiss, history helps to explain the causes and ramifications of these changes. As these insights from *Oceans Past* demonstrate, the HMAP approach is reducing the limits and conventional wisdom that restrict our knowledge and understanding of what used to live in the ocean. This, in turn, will help to generate the broad consensus and commitment required for the effective management and regulation of the marine environment.

REFERENCES

Ellis, R. (2005) *Singing Whales, Flying Squid, and Swimming Cucumbers: The Discovery of Marine Life*, Lyons Press, Guilford, CT

Halsall, P. (1996) 'The *Domesday Book* 1086', *Internet Medieval Sourcebook*, www.fordham.edu/halsall/source/domesday1.html, accessed 26 July 2007

Hilborn, R. C. (1994) *Chaos and Nonlinear Dynamics*, Oxford University Press, Oxford, UK

Hume, D. (1748) *Philosophical Essays Concerning Human Understanding: By the Author of the Essays Moral and Political*, London
O'Dor, R. K. (2003) *The Unknown Ocean: The Baseline Report of the Census of Marine Life Research Program*, Consortium for Oceanographic Research and Education, Washington, DC
Pratt, J. P. (1990) 'Yet another eclipse for Herod', *Planetarian*, vol 19, no 4, pp8–14
Waggoner, P. E. and Ausubel, J. H. (2002) 'A framework for sustainability science: A renovated IPAT identity', *Proceedings of the National Academy of Sciences*, vol 99, pp7860–7865, http://phe.rockefeller.edu/ImPACT/ImPACT.pdf

Index

abalone 66–71
  on menus 56, 68, 69–70
  prices 54, 61, 65, 66, 68, 70–71
acorn barnacle 12
affluence, and perceptions of fisheries 81
Alaska 191, 192
  Gulf of 135, 136, 143, 144, 146
alewives fisheries 77–78, 80–86
allelic diversity, *Littorina littorea* 14–15
Allison, C. and Smith, T. D. 151, 153, 159
allozyme analysis 13–14
Altham, Emmanuel 93
Alvin (submarine) 209
American Antiquarian Society 49
American lobster 61, 71–74
  see also lobsters
amplified fragment length polymorphisms 18
animal biomass and disease 43
Antarctic Ocean, sperm whale catch 159, 166
apathy and fisheries decline 86, 87
aquaculture 39, 66
archaeological evidence *see* fossil record
Arctic Bowhead fishery 136
Arctic Freezing Company 64
Argentina 191, 198
Arkhangelsk 180
Atlantic Ocean
  right whale fisheries 133
  sperm whale catch 159
  sperm whaling grounds 115, 116, 118, 119, 120, 122–123, 124, 126, 127, 128

  see also North Atlantic fisheries; Northwest Atlantic
Australia 191, 197
Aythya valisineria see canvas-back duck
Azov Sea 179

bacteria 42
Baer, Karl Ernst von 175–181, 183
Bahamas sponge fishery 28, 31, 40
Baird, Spencer Fullerton 193
bait fish 82–83, 84, 86
Balaena mysticetus see bowhead whales
ballast rocks 10, 12, 19
Baltic Sea 176–179
Bannister, J. L. 158
banquet menus 49, 50–51
Barents Sea 179, 180
Basque fishermen 93
Batillaria attramenteria (Asian horned snail) 15–16
Bay of Bengal 191
Beale, T. 110, 111, 113, 114
behaviour, sperm whales 152, 162, 163, 164, 165, 166–167
Belding, David 77, 83, 86
benthic invertebrates 43
Berger, E. 13, 14
Bering Sea 146
Best, P. B. 153, 154, 156–157, 159
Beverly, Massachusetts, logbooks 94, 97, 103, 104
Beverton, Raymond 202
bills of fare *see* menus
Black Sea 179
Blakeslee, A. M. H. 18–19

blue-fin tuna 65
bluefish 61
blue mussel 12
blue whales 109
Boeck, Jonas Axel 175, 182
Bohuslan fisheries 178, 181, 182
bowhead whales 135, 136, 146
Bristol Bay 191, 192
Bristol's Oyster and Chop House, New York City 59–60, 61
British Colombia 191
Bushman, Richard 79–80

Canada 109, 191, 192
    Littorina littorea 10, 11, 12
canning industry 191
    abalone 68
    lobster 71–72
canvas-back duck 61–65, 73, 74
Cape Cod
    alewife fishery 77–78, 80–81, 81–86
    inshore fishery 84
    offshore fleets 82
capitalist market structures and fisheries decline 78–79
Caracas, Tenth Inter-American Conference (1954) 197
Caribbean marine ecosystem 27, 42, 78
Caroline Islands 191
Caspian Sea 179–180
catch
    alewives 84
    Cape Cod inshore fisheries 84
    global 193
    historical records 96, 99, 100, 105
    marine sponges 30–31, 33, 35, 39, 42
    North Pacific right whale fishery 136, 143
    Russian fisheries 176–177, 179–180
    sperm whales 149–152, 153, 156–158, 159–161
censuses 207
Census of Marine Life (CoML) 1, 207, 208, 209, 212
chaos theory 209
Chapman, J. W. 15–16, 17–18

Chapman, Wilbert McLeod 192, 193–195, 197, 198, 200, 201
charts
    fishing 94–96, 97, 99, 100–101, 102, 106
    Maury's 133–134
    Townsend's 110, 112, 114, 129
Chesapeake Bay 8, 62
Chile 192, 196, 198
Chilean seabass 48, 65
Chinese fishermen 66–67
clams 52, 209
Clark, A. H. 110, 111, 113, 114–115, 119, 122, 123
Clark, Christopher 80
climate change 92, 153, 166
coastal weirs 83–84
cod
    Atlantic 189
    Barents Sea 80
    fisheries 78, 82, 93–94, 97, 99, 104–105
    prices 61
coelacanths 212
Cold War 189, 199
CoML (Census of Marine Life) 1, 207, 208, 209, 212
common periwinkle *see Littorina littorea* (marine snail or common periwinkle)
Comprehensive Ocean-Atmosphere Data Set (COADS) project 134
conservation
    abalone 66, 70
    alewife fisheries 81, 83
    methods 202
    migratory waterfowl 63–64
    sponge fisheries 37–40
    US policy on 198, 199–200
Consumer Price Index (CPI), normalized prices of menu items 56–60
consumers 47, 48, 74–75, 210
consumption
    abalone 66–71
    canvas-back duck 62–65
    of fish 47–48
    lobster 71–74

coral reefs 26, 41, 42–43
  deep sea 96, 100
corroboration 2
Costa Rica 192, 193, 198
CPI *see* Consumer Price Index (CPI)
crabs 52, 61, 209
Cryptocotyle lingua 18, 19
cryptogenic species 7–8
  Littorina littorea 8, 9–19
Cuba 30
Culinary Institute of America 49
customs house records 96, 154
  Arkhangelsk 180

Dago (island) 177, 178
Daily Catch Records (DCR) 99, 101–104
Danilevsky, Nikolay 179, 180
Davis, L. E. 154, 155
Dawson, J. W. 10
DCR (Daily Catch Records) 99, 101–104
digitization, fishing charts 99–100
disease 26, 44
  marine sponges 40–42, 43
dispersal mechanisms 8
  disease of marine sponges 41
  Littorina littorea 12–13
distant-water fleets 189, 190, 191, 192, 204
distribution of species
  historic data 103, 104
  sperm whales 115, 119, 162, 163, 164
divers
  abalone harvesting 66, 67
  sponge fisheries 25, 33–38, 39–40, 41
DNA sequencing 14–19
Doel, Ronald E. 190
Domesday Book 207–208, 209
Dominican Republic 198

ecological evidence, *Littorina littorea* 11, 12–13
economic competition, sponge fisheries 35, 36–38
Ecuador 196, 198
El Salvador 198
embryology 176

encounter rates
  right whales 135–136, 142, 143–146
  sperm whales 149, 150, 152, 162–165, 170–171
environmental constancy 152–153
equilibrium population dynamics 166
Eubalaena japonica (right whale) *see* North Pacific, right whale fishery
Europe, *Littorina littorea* populations 9–19
evidence of the past *see* charts; historical evidence; logbooks; Maury Abstracts; *Whalemen's Shipping List and Merchants' Transcript* (WSL)
Expedition for the Study of the Status of Fisheries in Russia (1851) 176–178
external markets 43
extinctions 26, 43, 189, 211–212, 213
  local 96, 100, 106
extreme events 210

Faeroe Islands 13
Falmouth, Massachusetts 81
FAO (Food and Agriculture Organization) 195, 196, 201
federal bounty regulations 99
Federal Geographic Data Committee (FGDC) 99
Finn, D. B. 201
fin whales 109
The Fish Gate (Graham) 201
fish hatching 84, 201
fishing agreements 91
fishing effort 104, 106
fishing fleets 82
  distant-water 189, 190, 191, 192, 204
fishing grounds
  Gulf of St Lawrence 102
  Northwest Atlantic and Gulf of Maine 95–96, 97, 100–101, 104
  see also whaling grounds
fishing records 91–99
fish populations
  early historical researches 175–184
  Norwest Atlantic and Gulf of Maine 91–94, 104–105

understanding of 202
Florida *see* sponge fisheries in Florida
Food and Agriculture Organization (FAO) 195, 196, 201
fossil records 8, 211–212
   Littorina littorea 11, 12
freedom of the seas 189, 192, 203
Fucus serratus 12
Fundy, Bay of 78, 95, 96
Future of Marine Animal Populations (FMAP) models 212

Gadus morhua (Atlantic cod) 189
   see also cod
Galápagos Ground 149–150, 152, 157, 158, 160, 163, 165
Galápagos Islands 191, 196
Gayetty, J. W. 67–68
genetic bottlenecks, *Littorina littorea* 8, 14
geographic information systems *see* GIS (geographic information systems)
geo-referencing 100
George's Bank 82, 96
German fishing regulations 181
giant squid 96
Gifford, Thomas 81
Gilmore, R. M. 110, 115, 119, 122
GIS (geographic information systems) 94, 99–100, 104, 106
global fish market 47–48
global food supply 193
Gloucester, Massachusetts 82
GMCP (Gulf of Maine Cod Project) 91
Goode, G. B., *The Fisheries and Fishery Industries of the United States* 94–96
Gosnold, Bartholomew 93
Graham, Michael 201–202
Great Depression 73
Greek divers *see* divers
Greenland 13
green turtle 52
Guayaquil Banks 196
Gulf Coast of Florida 33–36
Gulf of Maine Cod Project (GMCP) 91–92

haddock 50
halibut 50, 82, 91, 102, 191, 192
   prices 59–60, 61
Haliotis
   chracherodii (black abalone) 66, 71
   corrugata (pink abalone) 66
   fulgens (green abalone) 66
   rufescens (red abalone) 66
haplotypes, *Littorina littorea* 14, 15–16, 17
Harwich, Massachusetts 81, 83
Hawaiian Islands 136
hawksbill turtles 26
Hegarty, R. H. 153, 154, 157
Heincke, Friedrich 182
herring 84, 93
   Baltic fisheries 176–178, 181–183
Herrington, W. C. 195, 197, 198, 201
Hiiumaa *see* Dago (island)
Hippospongia 27, 28
historical evidence 2, 3, 213
   criteria-based 8
   fishing grounds 91–99, 101–106
   Littorina littorea 10–12
   menus 48–56, 74
   Russian fisheries 175–181
   Scandinavian fisheries 178, 181–183
   see also charts; logbooks; Maury Abstracts; *Whalemen's Shipping List and Merchants' Transcript* (WSL)
history 2–3, 176, 180, 212–213
   invisible 211–212
History of Marine Animal Populations (HMAP) 1, 87, 91, 208, 212–213
hoki 48
Holt, Sidney 202
Homerus americanus (American lobster) 71–74
Honduras 198
Hoplostethus atlanticus (orange roughy) 48, 65, 189, 210
hotel bills of fare 49, 52
humpback whales 109, 114, 166
hunting losses 157–158
hunting techniques, canvas-back duck 62, 63

hysteresis 210, 212–213

IATTC (Inter-American Tropical Tuna Commission) 190, 193, 195
Iceland 13, 192, 195–196
ICNAF (International Commission for North Atlantic Fisheries) 190, 193, 195
illegal markets, abalone 66, 71
illegal shooting 64
Ilyanassa obsoleta (eastern mud snail) 9, 10
IM (isolation with migration) analysis 18
Imperial Russian Academy of Sciences 175, 180
  archives 176, 177–178
Indian Ocean
  sperm whale catch 159
  sperm whaling grounds 113, 115, 117, 118, 125, 127, 128
inflation-adjusted retail seafood prices 60
Inter-American Tropical Tuna Commission (IATTC) 190, 193, 195
International Commission for North Atlantic Fisheries (ICNAF) 190, 193, 195
International Court of Justice 196
International Halibut Commission 198, 201
International Law Commission 195–196, 197, 200, 201
International Whaling Commission (IWC) 153, 158, 196
isolation with migration (IM) analysis 18
ITS sequences, *Littorina littorea* 14–15
IWC (International Whaling Commission) 153, 158, 196

Jackson, Jeremy 78
James Duncan Phillips Library 96
Japan 199, 203, 210
  distant-water fleets 191, 204
  relations with US 192
  territorial waters disputes 197
Japan Ground 149–150, 152, 157, 158, 160, 162, 163

Japan, Sea of 135, 136, 142, 143, 145
Johannesson, K. 13–14
Johnson and Wales Culinary Archives 49
Jonsson, Hannes 195
Judd, Richard 80

Kamchatka 191
  east coast 135, 136, 142, 143, 144
Key West 27–28
  sponge fishery 25, 28–33, 35–36, 38–39, 40
Knipowitsch, Nikolay 180
knowledge, limits to 208–213
Korea 191, 197
Krige, John 190
Kröyer, Henrik 182
Kurile Islands 135, 142, 143, 145, 191

Latvia 178
leasing fisheries 83
Library of Congress 49
Littorina
  littorea (marine snail or common periwinkle) 8–19
  obtusata 10, 13, 18, 19
  saxatilis (rough periwinkle snail) 9, 10, 13, 18, 19
Ljungman, Axel 175, 182–183
lobsters 26, 43
  on menus 50, 52, 53, 56, 71
  prices 52, 53, 58, 60, 61, 71–74
local extinctions 96, 100, 106
logbooks
  fishermen's 91–92, 94, 96–99, 101–103, 104
  oil yield 153, 156–157
  sperm whale catches 152, 158, 165
  Townsend Extracts 110, 126–127, 130
  see also Maury Abstracts
Lotze, H. and Milewski, I. 78
Lund, J. N. 153, 154, 155
Lyman, Theodore 85–86

McEvoy, Arthur 78
mackerel 61, 82, 93
Maine 71, 74, 80

Maine, Gulf of 92–93
  fishing charts 95–96, 97, 101, 106
Mariana Islands 191
marine resources
  changing local perceptions 77–78, 81, 84–85, 86
  historical records 91–99
  see also maximum sustained yield (MSY)
marine snail see Littorina littorea (marine snail or common periwinkle)
marine sponges 26–27
  disease 40–42
  see also sponge fisheries
Marshall, George 192
Marshall Islands 191
Marstons Mills Fishing Co 84
Massachusetts, alewife fishery 77–78, 80–86
'Materials for the history of fisheries in Russia and the seas belonging to it' (Baer) 180
Maury Abstracts 133, 134–136, 160, 162
  daily positions of whaling ships 137–141
  number of US whaling vessels 136, 142
Maury, M. F. 110–111, 133–134
maximum sustained yield (MSY) 192–195, 201, 202–203
  and US foreign policy 189–190, 200
MDIV analysis 16–17, 18
Melville, Herman 162, 163
menus
  as evidence 48–56, 74
  OPA statement 74
  see also prices
Mexico 191, 192, 193, 198, 199, 200
microbial communities, coral reefs 41–43
Mid-Atlantic Ridge 209
migration
  Bahamian sponge fishermen 28–30
  Greek divers 25, 33, 35, 38
  IM analysis 18
Migratory Bird Act (US) 1918 64, 65
migratory waterfowl, conservation 63–64
Mississippi Delta 62

mitochondrial DNA sequencing 14, 16–19
molecular evidence, Littorina littorea 11, 13–19
monastery archives 179, 181
Morris, S. R. 13, 14
Mytilus edulis (blue mussel) 12

Narva, Estonia 177, 178
National Archives (US) 49, 96
National Oceanic and Atmospheric Administration (NOAA)
  Comprehensive Ocean-Atmosphere Data Set (COADS) project 134
  database 60, 72
National Research Council (US) 105–106
native or non-native species 7–8
  see also cryptogenic species
New York, conservation legislation 63–64, 73
New York Historical Society 49
New York Public Library 49
Nilsson, Sven 178, 181–182
NOAA (National Oceanic and Atmospheric Administration) 60
North America
  European settlers 77, 79–80
  Littorina littorea populations 9–19
  prairies 61–62
North Atlantic fisheries 92–93, 195–196
North Pacific
  right whale fishery 133, 136–146
  sperm whale catches 166
  see also Japan Ground; Pacific Ocean
Northwest Atlantic 92–93
  fishing charts 95–96, 100
Norway 182, 192, 196, 199
Nova Scotia
  Littorina littorea populations 9, 10
  see also Scotian Shelf

Ocean Biogeographic Information System (OBIS) 212
Oceans Past: Multidisciplinary Perspectives on the History of Marine Animal Populations conference 1

Office of Price Administration (OPA) 73–74
oil landing data 151, 153–156, 158, 167–169
Onassis, Aristotle 197
On Herring and the Herring Fisheries (Boeck) 182
OPA (Office of Price Administration) 73–74
open-boat whaling 109–110, 128, 129, 150, 155, 165
orange roughy 48, 65, 189, 210
Organization of American States 197
over-fishing
   marine sponges 26, 32, 36–37, 41–42
   North Atlantic 189, 195
   oysters 43–44
   through leasing fisheries 83
oysters 26, 43
   on menus 50, 52

Pacific Ocean
   conflict over fisheries 191–192
   right whale fisheries 133, 135, 136–146
   salmon fisheries 78, 84, 190, 191
   sperm whale catch 159, 160
   sperm whaling grounds 112, 115, 116–117, 118–119, 125, 126–127, 128
Panama 198
Panama-Pacific Exhibition (San Francisco) 49, 70
parasites, species invasions 18, 19
path dependence 210
Peipus, Lake 176, 179
pelagic bacteria 42
Peru 191, 192, 196, 198, 199, 200
Physeter macrocephalus see sperm whales
pike 50, 178
Pinchot, Gifford 192
populations
   Littorina littorea 9–19
   sperm whales 149–150, 151–152, 158–160, 166, 167
   see also fish populations

prices
   abalone 54, 61, 65, 66, 68, 70–71
   bluefish 61
   canvas-back duck 61, 62, 64, 65, 73, 74
   cod 61
   crabs 61
   halibut 59–60, 61
   lobsters 52, 53, 58, 60, 61, 71–74
   mackerel 61
   marine sponges 35, 39, 41
   salmon 61
   sardines 61
   seafood 56, 60, 75
   shad 61
   smelts 61
private management of fisheries 83–84
processing killed whales 165
Progressive Movement (US) 192

race relations, south Florida 29–30
radiocarbon dating 12
reasoning from the parts to the whole 209–210
red-headed duck 64
refrigeration 63, 191
restaurant bills of fare 49, 53–55
Revel *see* Tallinn
Rich, W., *Fishing Grounds of the Gulf of Maine* 94, 95, 96, 97
Ricker, William 202
right whales 109, 133
   see also encounter rates; North Pacific, right whale fishery
river herring 82, 86
   see also alewives
Rochester, Massachusetts 81
Rome Conference 189, 197–198, 199–203, 204
Roosevelt, Theodore 195
Rosenberg, A. A. 78
Rosier, James 93
Russian fisheries 176–181
   see also Soviet Union

St Lawrence, Gulf of 102

St Petersburg 176, 177–178
Sakhalin 191, 197
Salem, Massachusetts 94
salmon 61, 84, 202
   Baltic fisheries 178
   Pacific fisheries 78, 84, 190, 191
Sandwich, Massachusetts 81, 83
San Francisco, abalone trade 67–68
San Francisco Bay 7, 62
Santiago Declaration 197
Sarah (whaler) 163
sardines 61, 202
scallops 52
Scammon, C. M. 110, 111, 113, 114
Scandinavian fisheries 178, 181–183
Schaefer, Milner B. 199–200, 202–203
Schulz, Alexander 178
science 211, 212
   role in changing attitudes 77–78, 85–86
   role in MSY 193–194, 199–200, 201–203, 204
   role in US foreign policy 190
Scotian Shelf
   biomass of cod 91–92
   fishing grounds 99, 103–104
   hook and line fishers 78
seasonal fishing areas 100, 101, 104
seasonal whaling areas 115, 119–122, 134, 144–145
Sebastes (west coast rockfish) 189
Semibalanus balanoides (acorn barnacle) 12
shad 50, 61, 84
Sherman, S. C. 154, 155
smelt 61, 176
Smith, Tim 202–203
source materials *see* historical evidence
South Pacific Fishery Conference 196
South Sakhalin 191
Soviet Union 191, 197, 199
   fishing fleet 190, 196
   see also Russian fisheries
species invasions 7, 19
sperm oil 151, 153–157
   see also oil landing data

sperm whales 111–112, 133, 149–153
   abundance 162–165
   populations 149–150, 151–152, 158–160, 166, 167
   removal estimates 153–158
   see also encounter rates; whaling grounds
spiny lobster 48
sponge fisheries in Florida 25–26, 42
   conservation ethics 38–40
   decline 40–44
   economic competition 36–38
   initial growth 28–33
   technological innovation 32, 33–36, 39, 43
'Sponger Money Never Done' 31
Spongia 27, 28
Starbuck, A. 153, 154, 155, 156, 157, 167–169
'stepping-stone' invasion 8, 13, 19
sturgeon 178
Sumaila, U. R. 56
surplus production theory 199, 202–203
Swedish fisheries 178, 182
swordfish 48, 65

Taiwan 191
Tallinn, Estonia 177, 178
Tarpon Springs, Florida 33, 35, 36, 39, 41
taxes 81, 178
Taylor, Joseph 78, 84
technological innovation
   New England offshore fleet 82
   refrigeration 63, 191
   sponge fisheries in Florida 32, 33–36, 39, 43
Tenth Inter-American Conference 197
terrapin 52
territorial waters, conflict over access 191, 192, 195–198
Tillman, M. F. and Breiwick, J. M. 149, 152, 158, 160
Townsend, C. H. 110, 111, 113–114, 119
Townsend extracts 110, 126–127, 130, 157
Townsend's Charts 110, 112, 114, 129

trematodes 15–16, 18, 19
Truman Proclamation (1945) 191, 192, 195
tub trawls 82
tuna 65, 202
  fisheries 190, 191, 192, 196, 203
  industry 192
turbot 94

United Kingdom (UK) 196, 199, 204
United Nations (UN) 195, 197
  see also Food and Agriculture Organization (FAO)
United States (US)
  distant-water fleets 191, 203, 204
  fisheries policy 192–195, 200
  foreign policy 189–190, 197–198, 199–200, 203–204
  political pressure from fishing industry 191, 192
unpredictability of humanity 210–211
US High Seas Fisheries Policy 192, 193, 194, 200

validated primary information 3
Vallisneria americana 62
vastness of the present state 208–209
Venezuela 197
Vikings 12
voyage based data 167–169

Wallerstein, Immanuel 79
Wares, J. 14–15, 16–17
water filtration by marine sponges 26–27, 41–42
water telescopes 32
weirs 83–84, 181

west coast rockfish 189
Western-style capitalism 79
whale encounters
  right whales 135–136, 142, 143–146
  sperm whales 149, 150, 152, 162–165, 170–171
Whalemen's Shipping List and Merchants' Transcript (*WSL*) 134, 136, 142, 143, 153, 154, 156, 167–169
whaling grounds 110, 122–128
  geographical boundaries 113–115, 116–118
  Japan Ground and Galápagos Ground 149–150, 152
  names of 111, 113, 128
  North Pacific 135, 136, 142, 143, 144, 145, 146
  seasonal distribution 115, 119–122
  whalers' information on 164
whaling industry decline 123, 127, 128
whaling vessels
  daily positions 134, 135, 136, 137–141
  number in North Pacific 136, 142
  see also encounter rates; open-boat whaling
Whitehead, H. 149, 150, 152, 158, 160, 162
White Sea 179, 180
Wilkes, C. 110, 111, 113, 114, 115, 119
William of Normandy 207–208
Willis, J. 10
Wolf, Eric 79
Wood, William 77
world systems theory 79
WSL see Whalemen's Shipping List and Merchants' Transcript